D1541439

THE SCIENCE
OF SUPERVILLAINS

Lois H. Gresh

Robert Weinberg

WILEY

John Wiley & Sons, Inc.

Published by John Wiley & Sons, Inc., Hoboken, New Jersey
Published simultaneously in Canada

For general information about our other products and services, please contact our Customer Care Department within the United States at (800) 762-2974, outside the United States at (317) 572-3993 or fax (317) 572-4002.

Wiley also publishes its books in a variety of electronic formats. Some content that appears in print may not be available in electronic books. For more information about Wiley products, visit our web site at www.wiley.com.

Library of Congress Cataloging-in-Publication Data:

Gresh, Lois H.
    The science of supervillains / Lois H. Gresh, Robert Weinberg.
      p.   cm.
    Includes bibliographical references and index.
    ISBN 0-471-48205-6 (cloth)
    1. Comic books, strips, etc. — History and criticism.   2. Villains in literature.   3. Science.   I. Weinberg, Robert E.   II. Title.
    PN6714.G75 2004
    741.5'09—dc22

                                                        2004003018

Printed in the United States of America

10 9 8 7 6 5 4 3 2 1

To my father, who gave his love of science fiction to me. And to my mother, with gratitude and deep love for everything.

—Lois H. Gresh

To Mark W. Powers and Pete Franco, two of the nicest guys ever to work in the comic book field.

—Robert Weinberg

On the Internet at:
www.sff.net/people/lgresh
and
www.robertweinberg.net

Contents

Preface

In the mid-1950s, comic books were going through a major slump. The most popular comics of the early 1950s, those featuring crime, war, and horror stories, had been swept off the newsstands by Senate hearings that tried to link comic books with juvenile delinquency. Though no direct correlation was ever proved, the bad publicity was so damaging that it forced comic publishers to invent a self-censoring code that all but wiped out violence, action, and death from their magazines. Humor and funny animal comics like the Disney brand sold well, but superheroes, long a mainstay of the industry, suffered.

DC Comics were kept afloat only by the iconic status of their two stars, Superman and Batman, whose popularity never wavered. Other comic book companies weren't so fortunate, and one after another, slid into bankruptcy during the 1950s. For a time, it seemed possible that superhero comics, an American mainstay since 1938, might perish. But two editors, Julius Schwartz at DC Comics and Stan Lee at Marvel Comics, each in his own way, turned the industry upside down. Superhero comic sales surged in the late 1950s through the early 1970s in what became known as the Silver Age of Comics. DC and Marvel became media giants, and the face of comic book publishing was changed forever.

Schwartz, a well-known science fiction fan and literary agent, had been working as an editor for DC since the 1940s. In early 1956, he was given the job of reviving interest in DC superheroes. His vehicle was a new comic, titled *Showcase*, which featured tryout stories for new superheroes. If a character sold well in its *Showcase* appearances, it was given its own comic. If sales were poor, the character was dropped.

The first character to appear in *Showcase* under Schwartz's directorship was the Flash. The character was a familiar one to Schwartz, who had edited a 1940s version of the hero for several years. This time, as editor of the series from the beginning, Schwartz decided to do things differently. A science fiction fan since the 1930s, Schwartz knew that readers liked stories that seemed authentic—that were based on some element of actual science. Even if the science was twisted, bent, and stretched to the limits, the factual circumstances of the story gave it a much more believable feel. And that, Schwartz felt, was the key to selling superheroes.

Thus police-scientist Barry Allen was dosed by a batch of electrically charged chemicals during a thunderstorm. His costume was made of recently developed miracle fibers, and every time he did some new and seemingly impossible feat, a footnote to the story noted that "Barry was able to run across water because he never broke the surface tension of the liquid." Schwartz went so far as to fill the blank spaces in his comic with science clippings and facts.

The first issue of the scientific Flash was a success, and Schwartz knew better than to gamble with success. With the introduction of each new Flash villain, an aside or cutaway revealed the scientific secrets behind the villain's incredible powers. Each time Barry Allen caught crooks using some astonishing scientific trick, Schwartz was sure to make it very clear how the stunt was performed. The real secret of the comic wasn't the actual science demonstrated but Schwartz's determination to keep the stories plausible. The adventures might not be scientifically accurate, but they seemed to make sense. And that was what mattered.

The publishing theory of the day was that if a formula worked once, it would work a dozen times. It's still considered true today. Following the Flash in *Showcase Comics* were the Challengers of the Unknown, Lois Lane, Green Lantern, the Atom, the Justice League of America, and many others, all with their own quasi-scientific backgrounds and all but a few earning their own comic books. It was the Schwartz formula of superheroes based on science that revolutionized DC Comics in the 1950s and 1960s.

Meanwhile, at much smaller Marvel Comics, Stan Lee, another longtime veteran of the comic business, was told by his publisher to create a team of superheroes to match the popularity of DC's newest sensation, the *Justice League*. Lee invented a group he named the *Fantastic Four*. Soon after, he came up with another superhero comic, the *Incredible Hulk*. And within a year, he added *The Amazing Spider-Man* to the Marvel roster. However, Lee didn't follow Schwartz's model of making his heroes scientifically plausible. Instead, he tried another idea new to comics. He made them into soap operas.

Marvel's success was based on the three elements that made soap operas on radio and TV so popular. First and foremost, Lee's characters, both heroes and villains, had personalities. They were complex individuals with likes and dislikes, good traits and bad. Plus, as dictated by story events, the personalities of his characters changed over time.

Second, Marvel Comics featured long stories, often filling the entire issue of the comic book. Most superhero and adventure comics at the time featured two or three stories per issue, making complex adventures impossible. Lee enjoyed writing complicated narratives, and soon his stories were stretching over two, three, or even four issues. Marvel Comics became serials.

Third, and perhaps the most radical change, Lee stressed continuity in his comics. His heroes learned from their mistakes, as did his villains. Events progressed in a continuous fashion. Characters changed and grew older, some even married, and others died. Continuity became important. Lee's superheroes had a history, a backstory that made their lives more compelling, more interesting.

Taken separately, Lee's three comic book changes weren't an entirely different method of storytelling. To a small extent, comics of the past had experimented with each concept. However, all three ideas combined gave Marvel comics a new look. For the first time, comic heroes such as the Fantastic Four had real lives and spent their time doing things other than fighting supervillains. The quartet even argued and displayed human emotions. Marvel Comics soon became known as the comics that took time to read, as compared to other comics that could be read in a few minutes.

For the next ten years, Marvel and DC Comics offered readers a distinct choice. Marvel had the continuing soap opera adventures, aimed more at teenagers than preteens, and filled with angst, emotion, and melodrama. DC pushed superheroes with a much more science fiction look and an air of plausibility in even their most impossible tales. Still, the successes of each company did not go unnoticed by the other. Plus, in the volatile comic book job market, where most writers and artists worked as freelancers not bound by long-term contracts, offers and counteroffers fueled a steady flow of talent from one office to another and back again. Increasingly, the products of the two giants of comic book publishing began to look the same. The end of the Silver Age of Comics was more the result of similarity than competition. Though rivalry between the two companies remained intense, the differences between their comics were slight.

If either company could be declared a winner in a struggle that saw both companies grow, it had to be Marvel. Stan Lee's brand of storytelling slowly engulfed the comic book world, and DC comic characters dropped from perfect to merely extremely good. Some of them even developed bad habits. Marvel, of course, in an effort to keep one step ahead of the competition, made their heroes even more dramatic and human, and their minor flaws mushroomed into personality disorders. Today, in the early twenty-first century, superheroes have evolved into mere mortals, with all the troubles and flaws of ordinary people. Assuming, of course, that ordinary people could leap tall buildings in a single bound or lift an army tank with one hand.

What of the veneer of scientific believability developed by Julius Schwartz for the DC line of superheroes? With the increased emphasis on characterization (that is, social and emotional problems), the emphasis on logical scientific explanations fell to the wayside. There wasn't enough space in a story to offer intelligent explanations for the superpowers used by the heroes and villains. Also, as continuity grew more complex, the constant return of old characters in new guises made scientific explanations of their talents repetitive. In trying to make comics more real for their audience, modern comic book writers sacrificed science in the name of plausibility.

It's our intention in writing this book to describe the comic characters who retain an aura of scientific believability and to explain why other characters are implausible. Because our previous book, *The Science of Superheroes*, dealt with some of the great comic superheroes, we felt it only fair in this one to examine some of the great supervillains. We think you'll be surprised to learn who turned out to be more plausible than the others, and we certainly hope you'll be entertained!

Introduction
by Chris Claremont

True story: Back in the day before PCs, e-mail, and the Internet, when advanced writing technology was an IBM Selectric and pretty much the entire comic book production process was done by hand, I was assigned the script on an issue of the *Incredible Hulk*. (For obsessives in the audience, we're talking *Hulk* #170, "Death from On High!" from a plot by Steve Englehart.)

So here's the sitch: Bruce Banner and Betty Ross Talbot are falling from roughly eight miles high. She's unconscious; he's having a serious anxiety attack. The writer, young, eager to prove himself, looking for a textual and emotional kick to make the circumstance viscerally exciting for the reader. Flash of inspiration—go to the mechanics of the moment. I rushed to my library, scrambling through the shelves until I came up with my old high school physics book, and—with the help of an old slide rule—calculated the acceleration of a falling body (lovingly explicated on page 2, panel 2). Wrote the scene, felt justly proud of myself, turned in the issue, moved on to the next assignment, very impressed with the Hulk that he could survive an impact of 11,000 miles per hour.

One teeny, tiny, irksome little problem. The Hulk was falling through Earth's atmosphere. Which brought into play a little fact of life called terminal velocity. You can only fall so fast, starting from zero, before air resistance puts a natural brake on your speed. No matter how far you fall, you won't fall any faster (for an adult, it's about 120 mph). Oops.

You'd be amazed how many people noticed. And called me on it, big time! Big life lesson for the young writer on the role of science in the gestation and presentation of stories.

Comics, especially modern mainstream (American) superhero comics, involve the art of the fantastic. We dream the most ridiculous and incredible things and, through a marvelous amalgam of story and art, bring them to reality. Characters who leap tall buildings in a single bound, bitten by radioactive spiders, yada yada yada.

The best thing about superheroes was that they always seemed to have a plausible rationale: Spider-Man had the radioactive spider-bite, DareDevil was kayoed by radioactive material, and the Hulk was irradiated by a gamma bomb (notice a trend here?). Even Superman hailed from a world under a red sun; under Earth's yellow sun and weaker gravity he suddenly sprouted all manner of exceptional physical abilities. In those days, explanations were more global and simple, echoing the anxieties of the time—the guy's an alien; the guy was exposed to nuclear radiation. Iron Man benefited from the latest buzzword in electronics: transistors. We writers gleefully exploited whatever aspect of science and technology provided the current hot button in the popular consciousness, taking it to as extreme a point as possible.

Same went for villains. For example, Magneto, whose name implied that he could control the forces of magnetism. What did that mean? He could manipulate ferrous metals. At one point, he could manipulate the iron content of the blood to establish mind control over others.

But when you think about it, that's just the tip of an incredibly monstrous iceberg because, you see, magnetism (or rather its Siamese twin, electromagnetism) is one of the four foundation forces of nature (gravity, electromagnetism, strong nuclear force, weak nuclear force). It is one of the bases of everything we comprehend to be energy and matter, or—going to extremes of hyperbole (because this is after all comic books)—existence itself. Twisting metal should be child's play; it's just the beginning. For Magneto, everything on Earth that involves electricity falls under his potential influence: power grids, communications, computers (hold a magnet up to a hard drive, then try to access your data—not a pretty sight). He holds the keystones of modern society in his hands.

Think of *The Matrix*. The ultimate weapon of the human resistance was an electromagnetic pulse generator; trigger it and the machines totally crash (makes you wonder why they didn't have more of those puppies stockpiled, but that's a whole different discussion, and it's movies anyway, and whoever said they had to make sense). Same applies to Magneto. Just by walking down the street, he could conceivably send human society cascading back to the preindustrial era. Why bother to manipulate iron in the blood? What is the human nervous system, the brain itself, but a self-contained bioelectric network?

Why couldn't Magneto manipulate that system to enhance it and make it more efficient (thereby amplifying the mutant abilities of his allies), or inhibit it, to degrade the performance of his adversaries? Why bother manipulating blood flow when you can strike at the heart of the central nervous system? He could conceivably destabilize the attractive forces that hold "matter" together, working on subatomic and quantum levels. I mean, might he even be capable of generating localized wormholes? Forty years he's been around, and the more we learn about the guy, the more we realize we've barely scratched the surface of his potentialities.

Herein lies the nature of what's happened to the craft of comics writing over the past generation. Y'see, writers are basically packrats (or to paraphrase the character of Sam Seaborn from *West Wing*, good writers borrow from the best; great writers steal outright). To create our work, we draw on every aspect of the world around us: people we know, situations we've been in, dialogue we've heard or spoken, rhythms of speech, nuances of behavior, gestures, idiosyncrasies, the works. All is gist for the creative mill. We who toil in the vineyards of periodical pulp fiction grab a lot harder and farther afield than most simply because the demands of our craft require us to produce work on a consistent and frequent basis. One standard series means a story every four weeks. Sure, the heroes are set (thank heaven for small favors), but that still requires the presentation of an adversary, either by creating someone new or bringing back someone who already exists, but in a way that showcases the character in a new and (somewhat) original light.

Like I said, back in the day, that didn't seem so much of a hassle. Cyborg a guy by stapling octopus arms to his spinal column, no problem. Give a guy extensible hydraulic legs that enable him to hopscotch over skyscrapers—why not? Give another guy functional bird's wings, totally cool. Invent an alien, absolutely.

Now, with the passage of time and the accretion of story, the bar of creation is continually set higher. Too often, when it comes to characters on either side of the line—hero or villain—we run into the syndrome of been there, done that. At the same time, however, the growth of our collective knowledge of science and technology has opened up huge new vistas of opportunity in terms of the powers we can come up with and the directions we choose. If transistors were cool, how much cooler might microbial nanites be? (Although I *still* want to know how Tony Stark fits those in-line roller skates inside his Iron Man boots, especially when the armor's folded up and tucked away in his briefcase, which when you think about it is the source of yet another paradox, because even if you assume some ability to condense the volume of the armor into such a containment, how do you deal with the integral mass? I mean, how much does that puppy weigh and how the hell does anyone this side of the governor of California carry it?)

In comics, which is as I said a descendant/variation of pulp fiction, you have the classic confrontation of hero and villain. Still (another paraphrase, I'm afraid, but you'll have to figure out this one for yourselves) the same old story, a fight for love and glory, a case of do or die. . . .

You want the hero to win, and for the most part you get your wish—that's axiomatic. But you also want a whole lot of collateral suspense, which means that the fight has to be a nail-biter, which means that the villain has to pretty much outclass the hero across the board. After all, imagine if Luke Skywalker were the size of Arnold Schwarzenegger. Would Darth Vader be as imposing an adversary? However, Mr. Schwarzenegger's very size made him the ultimate adversary against Linda Hamilton's Sarah Conner in *Terminator*. Since he looked and acted so utterly unstoppable, her final triumph against him was a true and lasting catharsis. Which is as it should be. The more terrifying and unstoppable the villain, the more satisfying the hero's victory.

But that means coming up with a really superior, kick-ass villain, and they're as hard to find as the names they go by. I look at the work on my desk right now and my catalog of adversaries runs the gamut from good to bad to indifferent: I have a telepath/telekinetic (doesn't everyone, and in the *X*-books, heartfelt sigh, doesn't one of them invariably have to be a red head?). I have an empath. I have a guy who can manipulate chains, make them move in all directions. I have superstrong guys. I have a gal who fires concussive blasts. I have a guy who can grow spikes out of his body, a gal who creates whirlwinds, another who makes the cellular membranes porous, one who throws quill-like darts, one who ties folks up in strands of silk. I have a character who can reshape the living flesh and bones of others as though they're clay. I have one with tentacles for arms.

Do I have the slightest idea how they work in real-world terms? Sometimes. I certainly do my best to try. That's why my desk is surrounded by bookcases, which are themselves crammed with all manner of reference books. To better comprehend what I can do with Storm, I read up on the weather. For Magneto, when he was active and under my pen, matter and energy and various realms of physics.

The problem is, so many characters, so much to look up, so little space to store it in. And no global index. So this is where I tip my hat to Bob and Lois, for boldly trekking where no star has gone before, on a mission to seek out new rationales for what makes these bad guys tick. Because, interestingly enough, as you ground them more and more in what may actually be possible, they become that much more plausible to both writer and reader and consequently more real. The writer gets to learn something new and interesting about the world and, through creative osmosis, try to pass some snippet of insight on to the readers.

I actually think this is a great idea of theirs that'll save me hours and hours of wandering around my office, or the Park Slope Barnes & Noble or the Brooklyn Public Library, searching desperately for that nugget of rationale that will lock the villain's powers and character into place. This way, I can gleefully thumb through this book and purloin the fruits of Bob and Lois's labors, tee-hee. And if I get it wrong, I can always blame *them!* Ain't life grand!

Gotta go now. I'm on deadline, with a moderately unbeatable bad guy about to do the terminal nasty to my heroes, and dispatching him involves the generation of a localized pinpoint singularity. (What's far worse is, once I lock him in this box, I have to figure out a plausible reason for his ultimate escape. Yeesh! Because this is comics, after all, where nobody dies forever!)

Enjoy!

1

The Original Dr. Evil
Lex Luthor

In **"How Luthor Met Superboy"** (*Adventure Comics* #271, 1960), Superboy flies to a farm in Smallville to introduce himself to a new kid in town. The kid, who has curly hair, is riding a tractor. A giant meteor of kryptonite falls from the sky and crashes to the ground next to Superboy. The farm kid whisks the meteor away with his tractor, depositing the kryptonite in quicksand. Superboy shakes the farm kid's hand, and the kid proclaims, "Meeting you, Superboy, is about the most thrilling thing that ever happened to me!" The curly-haired farm kid is Lex Luthor.

Luthor takes Superboy into his barn, where there is a shrine to Superboy: photos on the walls, rocks Superboy has punched, metal Superboy has bent. Luthor confesses, "I have hero-worshipped you for years. To me, you are the greatest boy in the world!"

The barn holds more surprises. It also contains an advanced scientific laboratory. As Luthor tells Superboy, his secret goal in life is to become a great scientist as famous as Superboy. To thank Luthor for saving him from the kryptonite meteor, Superboy builds an even more highly sophisticated experimental laboratory out of junk. He also gives Luthor rare chemicals, dug from far beneath the ground at superspeed.

For weeks, Luthor works to uncover the secret of life in his new laboratory. Finally, he creates a secret chemical formula that spawns a live protoplasmic monster.

Luthor's next goal is to create an antidote to kryptonite. He builds a giant claw arm and attaches it to his tractor, then lifts the giant kryptonite meteor from the quicksand. After chipping off some pieces, he drops the meteor back into the quicksand, then grinds the kryponite chunks into dust. He mixes the dust with the protoplasmic life-form that's still writhing—enormous hands flailing, head wobbling like jelly—in a big beaker on his laboratory table.

Unfortunately, as often happens in comics, Luthor stumbles and drops his beaker of protoplasmic monster plus kryptonite dust. The lab blazes with fire. Luthor chokes from the dust, fumes, and radiation.

Superboy arrives almost instantly and uses his superbreath to quell the flames. He rescues his friend, but this is a different Lex Luthor. He's been transformed by the radiation. He's bald, and he's insane. But he's still a scientific genius.

Luthor decides to destroy Superboy. He smashes everything in his shrine. He tricks Superboy into exposing himself to kryptonite in outer space, where Luthor maniacally chortles over the fact that Superboy's dog is choking half to death from the exposure.[1]

In later comics, Superboy often puts Luthor in prison. And Luthor always breaks out, using some weird scientific gizmo or technique. In one particular comic, Luthor is in prison and creates a salve that enables him to stretch his arm clear across Smalltown, where he creates havoc. There is an incredible image of Luthor with his elongated arm stretching across town from the jail cell. Luthor is indeed the villain of all comic book supervillains; he is the ultimate mad scientist.

Let's take a look at some of the scientific methods used by Lex Luthor and determine if they're at all plausible. In addition, let's ponder some techniques that Luthor never used and some he should have used, given his vast scientific expertise.

In the "farm lab" issue (#271) described above, Luthor creates a weather tower that transforms the sun's rays into solar energy in the depths of winter. Smallville installs the weather tower, and the town is blessed with summer flowers, crops, trees with leaves and buds, and gentle warm breezes. But then something goes wrong and the tower starts frying the town. Superboy saves the day, however, with his frosty breath.

Is it possible for solar energy to give Smallville summer weather in the depths of winter? Let's consider some facts. Of the solar energy reaching the Earth, approximately 30 percent is reflected and not used to heat the planet. The atmosphere absorbs an additional 20 percent of solar energy. The ground and oceans absorb the remaining half.[2] When it absorbs energy, the atmosphere grows warmer. The same thing happens with the ground and the oceans.

So how might Lex Luthor's weather control tower work? Our guess is that he uses sun charts and solar collectors. The position of the sun in the sky changes every day, and it differs depending on latitude (that is, where you are on the planet). However, for any time of any day, Luthor can predict the position of the sun in the sky. We figure that he then uses his mathematical predictions to design solar collectors that absorb the most sun and create the most heat.

Running on a computer, Luthor's mathematical predictions can take into account how Smallville's houses, schools, shops, trees, chunks of kryptonite, and other obstacles block the solar energy reaching his weather control tower. The math must also include calculations that compensate for fog, rain, dry spells, and pollution, as well as any blizzard, hurricane, or tornado that might hit the town.

Solar collectors are used today, of course. Even ten years ago, renewable energy supplied 18 percent of the world's energy.[3] Renewable energy either regenerates or cannot be depleted, and it includes solar energy.

Photovoltaic solar energy is used in calculators and wristwatches. It converts the sun's energy directly into electricity. Silicon combined with other materials in the photovoltaic cells enables electrons from the sun to move through the silicon, hence producing current. It does not appear that Lex Luthor uses photovoltaic power in his weather tower. Otherwise, he'd electrocute Smallville!

The weather tower is shown with a huge parabolic dish, which we assume collects and concentrates solar energy on a series of receivers. Once collected, this energy must be harnessed, but how?

Solar thermal systems transfer the sun's heat into fluids such as water. It most definitely does not appear that Luthor's weather tower converts the heat into fluid form. There is no evidence that Smallville is flooded by water or any other fluid.

While it's possible to collect and use solar energy, we do not believe it makes sense in the way that Luthor does it. One parabolic dish at the edge of town just doesn't suffice to radiate heat all over Smallville and its outlying farms, causing flowers to blossom in winter, crops to grow, trees to bud, and people to wear bikinis. That's just way too much heat being radiated from one dish. And we cannot conceive how that solar energy is being converted and distributed to such a wide area simply through the sky, with no liquid, no electrical equipment, and no conductors of any kind. Here we have a grounding in good science but an unfeasible application. At least according to current technology.

Most of Luthor's comic science leans heavily toward super-futuristic wonders. One example is teleportation. In "The Einstein Connection" (*Superman* #416, 1986), Lex Luthor perfects a teleportation machine and is able to make himself temporarily invisible whenever Superman gets too close. Superman follows Luthor to Princeton University. "You muscle-bound simpleton!" exclaims Luthor, as he zaps Superman with his "concussion blast" watch. Luthor steals a perpetual motion machine and creates illusions, such as multiple Luthors who don't really exist, rooms and walls that aren't really there, and waterfalls in the middle of nowhere.

Is teleportation possible? And is it possible to create illusions, such as multiple Luthors who don't really exist, rooms and walls that aren't really there, and so forth? A staple of science fiction, teleportation refers to the process of disintegrating a person or object in one place, then reconstituting the person or object in another place. It's generally done within seconds and usually never explained except in terms that are equally mystifying, such as *Star Trek*'s "pattern buffers."

Scientists have made some progress in the study of teleportation. Ten years ago, for example, a group of six scientists provided evidence that teleportation could happen, but only if the original person or object was destroyed in the process.[4] In 1998, using coaxial cable, scientists at the California Institute of Technology teleported a beam of photons from one end of a table to another. And in 2002, Australian physicist Ping Koy Lam teleported a laser beam approximately one meter from its origin by embedding a radio signal into the laser beam, destroying the original beam, then recreating it elsewhere. Although the original beam had to be destroyed during the teleportation process, the radio signal did survive.

The destruction is a result of the Heisenberg uncertainty principle, which states that the more definite you are about a particle's location, the less definite you are about its velocity, and vice versa—that is, you can't duplicate the exact spin and polarization of a particle at any given fragment of time. Instead, scientists use a process called quantum entanglement, whereby two photons are created simultaneously, and the changes made to one photon also occur in the other regardless of how far apart they are—for example, at opposite ends of the table.

Obviously, we are far away from teleporting humans. When you consider that a person—say, Lex Luthor—has approximately 10^{27} atoms in his body, it would be a great effort to teleport him a meter away, much less to another room, city, or planet. And once you teleport him, what happens to the original Lex Luthor? Is he destroyed due to quantum entanglement and the Heisenberg uncertainty principle? Are his exact thoughts and physical conditions (upset stomach, rapid heartbeat, infusion of narcotics into his blood, use of lifesaving heart medicines, and so forth) also transmitted? Probably not. The teleported Luthor would not be a perfect copy of the original Luthor.

Is it possible to create illusions, such as multiple Luthors who don't really exist, and rooms and walls that aren't really there? Two possibilities are that Luthor is using virtual reality or holographs.

Virtual reality is a computer-generated world in which we move and interact with objects, other real people, and virtual reality people. It's a place that doesn't exist but that offers the powerful illusion of tangibility.

Virtual reality today comes in two flavors. One surrounds you with three-dimensional objects and scenes so that you feel you are walking through a real place. This effect requires equipment: virtual reality goggles, for instance, or specially equipped rooms. Unless Superman is wearing goggles or other virtual reality equipment, Luthor is not using this form of virtual reality to produce his illusions.

The second type appears before you on a two-dimensional screen such as your computer monitor. The computer graphics and programming are so well done that a full three-dimensional world feels real on your two-dimensional screen. Many computer games are forms of virtual reality. Clearly, Luthor isn't using this type of virtual reality. Superman is not looking at Luthor on a computer screen.

So what is Luthor doing? Our best guess is that he's projecting holographs of walls, rooms, and himself. In holography, laser light is used to record the light-wave patterns reflected from an object or person. The light-wave patterns are placed into an emulsion of light-sensitive film. After the film is developed, it is again exposed to laser light. All points of light originally reflected from the object or person are captured, and the final image—whether you're standing in front of the holograph, behind it, or to its side—looks just like the object or person.

If Superman tried to touch one of the Luthor holographs, his hand would find only light. In a similar fashion, if Superman tried to punch his way through a wall, he'd be banging through empty space.

Luthor's holograph machine can work in various ways. For example, if Luthor is using a reflection holograph technique, then he's lighting his illusions from the front. If he is using a transmission holograph technique, then he's lighting his illusions from the back.

Most likely, he's employing pulsed holography or integral holography. Pulsed holography uses bursts of laser light to record a subject's movements. If Luthor's illusions move a good deal, then it's more likely that he's using integral holography, which converts individual frames of two-dimensional film footage, computer graphs, or digital video into holographs.

In another deadly scene, "The Luthor Nobody Knows" (*Superman* #292, 1975), Luthor escapes from prison and takes over a helicopter on an army base where they're detonating nuclear bombs. He steals a nuclear bomb and flies very close to a town. As Superman rescues Luthor from the helicopter, it explodes, and the nuclear bomb goes sky high. Somehow, the people in the nearby town are not hurt.

Now if a nuclear bomb explodes about a mile outside town, would the people be hurt? And what other methods might Luthor use to kill people from his helicopter?

Although it's true that bombs existed in 1975, their use in the Lex Luthor comics was more than fantastic. It seems that the Luthor bombs were not based on scientific fact.

Although battlefield nukes have existed since the mid-1950s, this is not the type of nuclear weapon that Lex Luthor uses near the town. An example of a battlefield nuke is an M573 or M422 8-inch nuclear projectile. Providing on-ground attack capability, these nukes tend to hit their targets with great accuracy. During the 1990s, however, the army and marines replaced most on-ground nuclear artillery weapons with more conventional weapons.[5]

But Luthor's bomb was the type dropped on Japan during World War II—a free-fall bomb. He dropped it from an airplane. He did not load it into a rocket launcher outside of town or from a submarine deep beneath the surface of the Pacific.

Clearly, if such a nuclear bomb were to explode about a mile outside town, most of the townspeople would die. World War II's Fat Man Model 1561, for example, destroyed 1.5 square miles of Nagasaki and killed 35,000 people. The Little Boy bomb destroyed 4 square miles of Hiroshima and killed 70,000 people. When accidentally dropped in 1957, a Mark-17 bomb created a 25-foot-wide, 12-foot-deep crater and threw debris to locations over a mile away. During this accident, only the bomb's conventional explosives detonated—imagine the destruction had the nuclear arsenal detonated.

In general terms, a nuclear bomb releases the forces that hold the nucleus of an unstable atom together. This can be accomplished in two ways. With nuclear fission, the nucleus is split into two fragments; isotopes of uranium or plutonium are typically used. With nuclear fusion, two atoms are brought together; hydrogen or hydrogen isotopes are typically used.

There are many ways of devising and detonating bombs. Some of the most common nuclear bomb designs are

- Fission bombs (the earliest type of bomb)
- Gun-triggered fission bombs
- Implosion-triggered fission bombs
- Fusion bombs

To understand how a fission bomb works, you need some basic knowledge about nuclear radiation. Each atom consists of subatomic particles: protons and neutrons form the atom's nucleus and electrons orbit the nucleus. Protons have positive charges, electrons negative charges, and neutrons no charge at all. Usually, the numbers of protons and electrons in an atom are the same. The role of the neutrons is basically to keep the protons together in the nucleus. Because the protons all have the same charge—positive—they repel one another.

Some elements have more than one stable form. By stable, we mean that you could leave the element alone for five hundred years, then return to find that it hasn't changed at all. For example, the copper, gold, and silver in objects found during the excavation of an ancient Roman city are exactly the same as they were two thousand years ago.

In fact, speaking of copper, 70 percent of all natural copper is called copper-63, and the other 30 percent is called copper-65. Each type of copper has 29 protons, but a copper-63 atom has 34 neutrons and a copper-65 has 36 neutrons: similar, but slightly different. Both copper-63 and copper-65 are stable forms of the element. Both are called isotopes of copper.

Some isotopes happen to be radioactive. In the most simple terms, radioactivity means that an isotope is unstable. For example, one of the hydrogen isotopes, which is called tritium, is radioactive. It has one proton and two neutrons. Over time, it transforms by means of radioactive decay into the more stable isotope called helium-3, which has two protons and one neutron.

Pretty cool, huh? Now there are three ways that a radioactive isotope will decay: alpha decay, beta decay, and what we're interested in talking about here, spontaneous fission. (This is how, by the way, alpha, beta, gamma, and neutron rays are formed.) For example, a fermium-256 atom, which is really heavy, may split into one xenon-140 atom and one palladium-112 atom, and in the process, shed four neutrons. These four neutrons may crash into other atoms and cause various nuclear reactions.

Induced fission means that an element can be forced to split. Uranium-235, often used in fission bombs, is a good example of such an element. If a uranium-235 nucleus is hit by a free-floating neutron, then the nucleus instantly becomes unstable and splits. This kind of thing happens to cause a nuclear explosion.

In a gun-triggered fission bomb, explosives propel a uranium-235 bullet down a barrel. The bullet hits a generator, which launches the fission reaction. Basically, Little Boy held two masses of uranium-235 nuclear material at each end of a tube, and when an explosive charge fired from one end, it shot nuclear material toward the other end. With all of the nuclear material essentially combined into one explosive force, a nuclear chain reaction occurred and released enormous energy. This energy caused a massive explosion. Detonated over Hiroshima, Japan, during World War II, Little Boy was a gun-triggered fission bomb. It had a yield equal to 14,500 tons of TNT and leveled most buildings within 4 square miles of ground zero.

In an implosion-triggered fission bomb, explosives create a shock wave that compresses the core of the bomb. The fission reaction occurs, and the bomb explodes. Fat Man, which devastated Nagasaki, was an implosion-triggered fission bomb. It contained a

13.5-pound sphere of uranium-235 and plutonium-239 surrounded by explosives. When the explosives fired, shock waves compressed the plutonium, increasing its density by two. At this point, a nuclear chain reaction occurred, causing the bomb to explode.

Which brings us to fusion bombs, also known as thermonuclear bombs. These are far more powerful than either Little Boy or Fat Boy. Basically, the fission part of the bomb implodes, and resulting X-rays heat the inside of the bomb. Pressure causes shock waves that initiate the fission in a plutonium rod, which in turn gives off radiation, heat, and neutrons. Combined with high pressure and temperature, these neutrons are used to create fusion reactions, which produce even more radiation, heat, and neutrons. In a horrific cycle, the neutrons from the fusion create more fission, and round and round we go until the bomb detonates.

Even if Luthor dropped his bomb farther from Smallville, irreversible damage would occur. This damage would be in the form of (1) intense heat and fire, (2) intense pressure, (3) radiation, and (4) radioactive fallout. The fallout alone would enter the water, cling to the air, and be carried to far distances by winds.

A 1-megaton hydrogen bomb possesses 80 times the deadly power of 1945's Big Boy. Within a 1.7-mile radius of its ground zero, everything would be destroyed, including 98 percent of the people.[6] Within a 2.7-mile radius, everything would be destroyed, including 50 percent of the people, with 40 percent of the remaining population seriously injured. Moving to a 4.7-mile radius, most buildings would be destroyed, with 5 percent of the people dead and an additional 45 percent of the population seriously injured.

So what kind of bomb does Lex Luthor drop a mile outside of town? A stink bomb? It is impossible that Luthor drops a nuclear bomb outside of town and nothing much happens in Smallville.

If Lex Luthor is determined to kill everybody, we wonder why he doesn't use biological and chemical weapons. For example, a modern Luther would contaminate water or air with anthrax, smallpox, ebola, or other deadly biological diseases. Anthrax is a bacterium that was used during the 2001–2002 terrorist attacks on the

United States. Smallpox is a virus that is highly contagious and spreads quickly through the air. There is no cure for ebola, an extremely lethal virus that induces massive bleeding in its victims. There's also no cure for Marburg, another extremely lethal virus that causes hemorrhagic fevers. Botulism, a bacterium, can be inhaled or ingested; it causes paralysis and respiratory malfunction. Any of these biological weapons in the hands of an insane super-genius like Lex Luthor would mean death not only to all the people in Smallville but to those throughout the country, the world, and given the life-forms on other worlds in the Superman series, death possibly to aliens whose biology makes them susceptible.

The same is true for chemical weapons. If Lex Luthor really wants to kill everyone, then he should use nerve agents, toxins, and mustard agents. Basically, chemical weapons are lethal gases and liquids that attack the lungs, blood, skin, or nervous system. They are stockpiled all over the world. For example, one estimate places the Soviet supply at 40,000 tons.[7]

Luthor can pick from blister, vomit, choking, blood, mentally incapacitating, nerve, and tear agents. Or he can mix batches of chemical weapons that include varieties of deadly agents. Typically, medical symptoms occur within minutes or hours. Smallville would be decimated—poof, in no time at all.

But Lex Luthor is far more than a modern homicidal lunatic. According to "The Impossible Mission" (*Superboy* #85, December 1960), Superboy returns to the day in 1865 when Abraham Lincoln was assassinated. And whom should he find there? Why, none other than Lex Luthor! We later learn that Luthor has invented a time-traveling machine to escape—from prison, no doubt—into the past. After Lincoln is shot, Luthor returns to modern time, wracked with guilt, feeling responsible for Lincoln's murder. But is it possible to travel backward in time? As with the case of teleportation, the question of time travel is an old staple of science fiction, stretching back beyond H. G. Wells's 1895 novel, *The Time Machine*. In 1888, Edward Page Mitchell published a story called "The Clock That Went Backward," but few people remember him, whereas Wells's story will always be famous.

Today, it is almost universally believed that Wells's time machine could not be built as easily as he presumed. However, leading scientists are willing to admit the possibility, at least on a theoretical level, that backward time travel may someday be possible. For example, physics and astronomy professor Jeffrey Kuhn comments that "the notion you can move forward and back in time is allowed by some of the new ideas in physics."[8] And in the August 2003 issue of *Wired*, noted professor of theoretical physics Michio Kaku writes, "Once confined to fantasy and science fiction, time travel is now simply an engineering problem."[9]

Scientific theories about time travel are based on quantum mechanics and Einstein's theory of relativity. It was Einstein who first proposed time as another dimension of physical reality, with the speed of light being the absolute speed limit throughout the universe. Evolving notions of space-time pushed scientists toward pondering time travel via black holes, cosmic strings, and space-time warpage.

Think of space-time as a large piece of cloth held at each of four corners. An object—matter—drops on the cloth, and the cloth sags: it warps, it curves. In this case, matter induces space to curve. Now a second object—matter—drops on the cloth, and this second object rolls toward the curve made by the first object. In this case, curved space induces matter to move.

Now imagine that a huge object falls on the cloth. In fact, the object is so huge that it rips the cloth: the fabric of space-time is now torn. Think of the rip as a black hole.

In outer space, as large stars deplete their nuclear fuel, they shrink and become increasingly dense. This denseness increases the star's gravitational pull. Eventually, the gravitational pull is so great that nothing can withdraw from it; not even light can escape. Indeed, the gravitational pull is so tremendous that it causes space-time to warp and tear—just as the huge object dropped on our cloth caused it to warp and tear. This is a real black hole.

It is believed that time slows in a black hole. It is also generally theorized that somewhere within the black hole lies infinite density, where the laws of quantum mechanics are no longer valid. Anything entering a black hole is sucked into an abyss and obliterated. Anything coursing along the edges of a black hole—far enough from its center to survive—could very well enter time travel.

But keep in mind that this form of time travel may be to another region of this particular space-time, to another space-time, or even to another universe several space-times away. If you're cruising in a spaceship around the rim of a black hole and you enter time travel, it is unlikely that you can specify such a thing as "take me to the day President Lincoln was killed."

Kip Thorne, Feynman professor of theoretical physics at Caltech, explains that if you tried to travel back in time to change your parents' destiny, both you and the time machine would be fried.[10] His answer, however, is tongue in cheek. Professor Thorne, who lectures and writes frequently about time travel and black holes, speculates that at the moment the time machine was activated, it would be destroyed by an intense beam of radiation composed of "vacuum fluctuations of quantum fields." He bases this speculation on the laws of quantum fields in curved space-time. Yet he further says that right before the time machine was destroyed, the laws of quantum gravity would replace the laws of fields in curved space-time, that we know very little about the laws of quantum gravity, and that within this vast bank of unknown knowledge lies the answer to the big question: Can we travel backward in time?

Many scientists suggest that the warp/tear in space-time caused by a black hole actually spews energy into another part of the space-time fabric—that is, to another region of this particular space-time, to another space-time, or even to another universe several space-times away. This event is called a wormhole. If you drop into a wormhole and somehow survive, you theoretically could end up in an entirely different dimension. And if all universes and space-times are interconnected by wormholes, it'd be as if you dropped into the plumbing of your house: winding your way through a labyrinth of pipes connecting all these space-times until finally you emerged and remained stationary. At this point, you could possibly be on Earth in America on the day Lincoln was killed.

Of course, if time travel is indeed possible, and if we can travel from the future to the present, then why haven't we encountered any of our descendants? Nor have any of them contacted us. At least we don't think they have.

We could fill an entire book describing Lex Luthor and his scientific marvels. In parting, however, we should mention one of the nice things he tried to do. In "The Last Days of Ma and Pa Kent" (*Superman* #161, May 1963), the Kents become very ill and may die. The warden releases Lex Luthor from prison temporarily so that he can attempt to cure Ma and Pa Kent. He has invented a vibro-health restorer that destroys all symptoms of diseases and cures all illnesses instantly. Not only can Lex Luthor devise and use a dozen futuristic methods of death in one day, he can also invent cures for fatal diseases. Unfortunately, however, Luthor's machine does not work. And after only one hour of trying to be a good guy, Lex Luthor returns to prison. Is it a coincidence that in this one instance real science has its day?

2

The Villain in the Iron Mask
Dr. Doom

According to most Marvel Comics handbooks and readers' guides, the most intelligent scientist in the Marvel universe is Reed Richards, leader of the Fantastic Four. He has invented all sorts of miraculous devices, such as superrocket fuel, a roving-eye TV, the incredible Fantasti-Car, the Energy-Jammer, and the Counter Sonic-Harness. So why didn't he make the effort to shield his original spaceship from cosmic rays, which exposed him and his companions to deadly energy bursts that changed them into the Fantastic Four? The answer lies in the words of his girlfriend and later wife, Sue Storm, who told test pilot Ben Grimm before their dangerous voyage, "Ben, we've *got* to take that chance . . . unless we want the Commies to beat us to it."[1]

With the collapse of the Soviet Union, such concerns seem ridiculous, but they were very strong motivating forces back in the days of the Fantastic Four's earliest adventures. Still, while the Fantastic Four sometimes fought Russian or Communist Chinese villains, most often they fought supervillains with few ties to politics. The most memorable of these characters had to be Reed Richards's personal nemesis and rival superscientist, Victor von Doom.

Victor von Doom was born in Latveria, one of those small European countries not far from Graustark and Freedonia. His father was a gypsy healer named Werner von Doom (perhaps a distant relative of the famous German rocket scientist?) and a gypsy witch woman, Cynthia. They both died when Victor was a child, and he was raised by his father's friend, Boris. Despite being educated by gypsies and roaming the countryside instead of attending public school, Victor became a scientific genius. Taking after his mother, he also became a master of the black arts.

In a moment of synchronicity common only to comic books and movie scripts, Richards and von Doom attended State University at the same time. Despite never having attended grade school, von Doom's reputation as a brilliant student reached America, and State University offered him a scholarship. Doom and Richards became scientific rivals. Anxious to make contact with his dead mother for reasons never stated, von Doom built an interdimensional communication device to bridge the gap between the living and the dead. Unfortunately, his design was faulty. Richards tried to inform von Doom of his mistakes, but von Doom ignored the warning. His machine exploded, horribly scarring his face. Because devices that communicated with the dead were illegal on campus, von Doom was expelled. He blamed Richards for sabotaging his work and instigating his dismissal, and the two became bitter enemies.[2]

von Doom traveled to Tibet, where a secret order of evil monks helped him forge special body armor along with an iron mask to hide his scarred face. In his rush to put on the iron mask, von Doom donned the faceplate before the metal was completely cool, thus damaging his face even further.

While Richards completed work on his Ph.D. at State University, von Doom majored in superscience and demonic studies. It's interesting to note that Reed Richards rarely if ever uses the title doctor, whereas Victor von Doom always does.

von Doom achieved his revenge in Latveria by overthrowing the tyrannical ruler of the tiny country and installing himself as absolute dictator. As ruler of a sovereign nation, von Doom possessed diplomatic immunity, which enabled him to bedevil the Fantastic Four without fear of retribution from local law enforcement or the U.S. government.

For forty years, Victor von Doom has been Reed Richards's—and by default, the Fantastic Four's—worst nightmare. He is relentless and unforgiving. His will, like the grotesque mask he wears to hide his horribly mutilated face, is as strong as iron. In the Marvel universe, Dr. von Doom, the madman in the iron mask, is the most dangerous supervillain alive.

Dr. von Doom's costume has remained unchanged for over forty years, which even by comic book standards is a long time. For his outer garments he wears a forest green jerkin with a black belt, a matching green hood, and a green cloak. His iron mask is a grisly parody of a human face, with rectangular holes for his eyes, a noseplate, and a metal mouthpiece constructed in the shape of a metal grin. To the world at large, the madman called Victor von Doom always appears to be laughing. But it's von Doom's armor, which keeps him alive even under the most dire circumstances. Leaving us to wonder, what does a non-superpowered supervillain wear?

Body armor has existed since the beginning of human history. For thousands of years, every advance in weaponry brought an equal advance in personal armor until the invention of guns and cannons in the sixteenth century changed the world. It wasn't until the mid-twentieth century that armor managed to catch up with weaponry, but just barely.

Bulletproof vests provide the user with some protection against gunfire, but all too often, not enough. A modern bulletproof vest doesn't use metal but high-tech woven fibers to protect the wearer. Such vests are called soft body armor, which is obviously what von Doom wears underneath his iron-plated armor.

Soft body armor is based on the principle of spreading the energy at the point of impact of a bullet (or other missile) over a wide area, thus lessening the blunt trauma caused by the projectile. This dispersal occurs using an interlaced net of anchored tethers that form an interlocking pattern to absorb the energy no matter where the projectile hits. In most bulletproof clothing, long, thin strands of Kevlar fiber make up the netting. Kevlar is a lightweight fiber made by DuPont that is five times stronger than a strand of steel. When thickly woven, Kevlar is extremely dense and almost impossible for a regular bullet to penetrate.

The momentum from a bullet is often powerful enough to break bones, which is why bulletproof vests are usually made from several layers of Kevlar netting and plastic film. To increase the protection offered by soft body armor, ceramic and metal plates are often inserted in pockets in the front of the armor. Unfortunately, no soft body armor is 100 percent effective. Even Dr. von Doom's combination of steel armor and soft body armor would not stop a barrage of tungsten core bullets. Armor dense enough to protect von Doom from such an assault would weigh so much that he'd be unable to move. Unless—

Unless his mysterious armored suit was something more than mere body armor. This is a distinct possibility. More than once in his battles against the Fantastic Four, von Doom (or more precisely, his armor) has demonstrated amazing powers, from being able to lift huge weights to navigating in outer space without any explanation. It could be that von Doom's primitive-looking body armor is actually a very advanced human exoskeleton suit.

In simple terms, a human exoskeleton suit consists of a robotic-type device that can be strapped on or attached directly to the human body. The device adds muscle power for heavy lifting, long-range running, and walking. It also enables the user to wear heavy armor without being affected by the weight. Such machinery has been common in Marvel Comics for decades, being the favorite uniform of many mercenaries working for outlaw political groups such as AIM and HYDRA.

Exoskeletons are even more popular in Japanese manga (comics) and anime (animated TV shows and movies often based on manga). Exoskeletons played a major role in the third Matrix film, *Revolutions*, and an exoskeleton that was normally used as a cargo loader helped Ripley save the day in *Aliens*. After decades of being promoted by pop culture, comic books, and cartoons, exoskeletons are now on the verge of becoming reality, especially if the U.S. government has anything to say about it.

In January 2001, the U.S. Defense Advance Research Projects Agency (DARPA) awarded approximately $50 million in contracts to laboratories and experimental groups to develop technology aimed at building an exoskeleton suit for ground troops. Dr. Ephrahim Garcia, coordinator of the project, said that the goals of the program are "formidable" and that "there is a huge challenge here." Dr. Garcia made it clear that the exoskeleton suit had to be something that soldiers could wear and use without thinking, not something controlled by switches.[3]

The exoskeleton would have to be able to perform specific tasks listed by DARPA that would allow troops to do the following:

- Increase strength—for example, carry heavier packs (including body armor), lift heavy objects, and use larger weapons
- Increase speed—for example, march faster over longer distances
- Leap extraordinary heights and distances

The first exoskeleton requirement would enable soldiers to carry large weapons into battle. At present, soldiers carry a backpack that is no more than one-third of their weight into war zones and often much less. It's common for them to leave behind equipment that is too heavy to carry for long amounts of time. The extra power requirement would also enable the exoskeletons to carry up to 10 pounds of extra protective gear for the user, not counting the armor on the exoskeleton itself.

Early work sponsored by DARPA involved pneumatic muscles or deformable magnets to power artificial limbs or suits that soldiers could wear. Pneumatic muscles were first invented in the early 1950s by physicist J. L. McKibben to help polio patients. These muscles were similar to balloons, which acted like pneumatic springs when put under pressure. The correct pressure in the balloon was maintained by a gas cartridge that produced carbon dioxide. At the Man Machine Systems and Control Department at the University of Technology in Delft, Holland, scientist Richard van der Line used pneumatic muscles to construct a walking robot named BAPS (Biped with Adjustable Pneumatic Springs).[4] In the United States, the Springwalker system from Applied Motion Inc. that was developed for the exoskeleton research project propelled its user at speeds greater than 15 miles per hour.

Other requirements for the exoskeletons would be that they would not require refueling for at least twenty-four hours and that they would move silently. The exoskeletons would also include a sensor web, expanding the user's field of vision, and would use thermal cameras to relay information about the battlefield to the wearer. Groups of soldiers wearing exoskeletons would be connected by global satellite positioning systems, enabling them to track each other in any situation.[5]

Are these goals unrealistic? The officials at DARPA and the scientists already working on test projects don't think so. In 1965, General Electric Research and Development Center, working with the U.S. military, developed an exoskeleton powered by hydraulics and electricity that they called Hardiman, which made lifting 250 pounds feel like 10 pounds. Unfortunately, the inventors of the robot, which weighed several tons, could only get one arm of the machine to work at a time.[6]

Recent results have been more promising. Researchers at Oak Ridge National Laboratory in Tennessee have invented a machine that can amplify hand motions to move heavy objects with ease and precision. This lifter can raise a 1,000-pound bomb as easily as a carton of cola.[7]

Scientists at the University of California at Berkeley's human engineering laboratory have constructed a highly advanced motorized exoskeleton to help disabled people walk. The exoskeleton weighs as much as an average adult and is powered by a chainsaw engine. However, when attached to a researcher's back and legs, it supports him as he walks, with the weight of the machinery completely unnoticed.[8]

Exoskeletons are coming, perhaps sooner than we realize. The walking device is not far from being used by disabled people. Long-term plans call for complete human exoskeletons by 2010. The cost per unit is projected at about the same cost—approximately $7,000—as a motorcycle. The face of warfare is about to change.

•   •   •

Sometimes comic books actually do predict the future. In the case of exoskeletons, they appear to be right on target. The Marvel universe is a reflection of our world—if only superheroes and supervillains, space aliens and mutants, monsters and mythological beings existed. It's not difficult to imagine that a young von Doom was an early pioneer in exoskeleton research. Finding financing for such work wouldn't have been difficult, as the governments and big corporations of the Marvel universe are always willing to pay big bucks for engineering breakthroughs. It's quite possible that someday a story in *Fantastic Four* might reveal how Dr. von Doom's specially constructed armored suit was financed by the CIA as part of a plot to overthrow the tyrannical ruler of Latveria. And thus how our own government was responsible for the rise of the most dangerous villain in the entire Marvel universe—Dr. von Doom.

3

Computer Supervillain or Village Idiot?
Brainiac

Brainiac is one of the few supervillains who is not based on a human life-form. Whereas most supervillains we cover in this book were humans before terrible lab accidents (most involving radiation and chemical spills) turned them into evil mutations, Brainiac has always been a computer who has a brain for data storage—an alien who came to Earth, although he does live in the body of a man.

Brainiac represents the early uneasiness about computers. In the 1940s and 1950s, many people worried that eventually computers would conquer humans simply due to their vast intelligence. There was even a great fear that computer software would be downloaded into the human brain, a concept that remains a staple in science fiction today. In Brainiac's origin story ("The Super-Duel in Space," *Action Comics* #242, 1958), Clark Kent and Lois Lane are on the first manned rocketship zooming through outer space. A flying saucer zips past the reporters' rocketship, shooting rays at it. Using his X-ray vision, Superman sees Brainiac for the first time.

As Superman watches helplessly, Brainiac sucks the Earth's largest cities into bottles on his flying saucer. As his saucer recharges its batteries using cosmic ray power, Brainiac lands on a nearby planetoid and battles Superman. Of course, Superman eventually outwits Brainiac and saves the Earth's cities, as well as Kandor, a city on Superman's home planet of Krypton.

In subsequent comics, we learn that Brainiac was designed by evil geniuses on the distant planet Colu to look like themselves: humanoids with green eyes and light green skin. Later in the series, we find out that Brainiac can successfully masquerade as a human (except for the fact that he is green with electrodes on his head) because his creators also endowed him with the mental patterns of a great Colu scientist. They created him to help them conquer Colu, the first step in their plan to take over the universe. But much later, a revolution on Colu destroyed them. Only their great achievement, Brainiac, survived.

Future comics put a new twist on the original story. For example, in *Adventures of Superman* #438, Clark Kent and Lois Lane go to the circus and see a psychic named the Amazing Brainiac, who is really a middle-aged sad sack named Milton Moses Fine. Fine suffers from horrible headaches, alcoholism, and a sharp-tongued common-law wife. He later tells Clark that he is the scientist Vril Dox from the planet Colu; further, that his green body was destroyed and only his brain survived. Apparently, Brainiac is inside the head of Milton Moses Fine.

So at this point, it seems that Brainiac's essence is as a computer mind only and that his original green-skinned android body is simply a nesting place. Brainiac is hiding within Milton Moses Fine so that Earth people will think he is a fellow human.

Later, in *Adventures of Superman* #445, Fine's body tries to expel Vril Dox's brain, and in the process, Fine gets a brain tumor and loses his outer layer of skin, revealing his true green skin beneath.[1] So, suddenly, the green skin is back. Apparently, Brainiac's computer mind affects the skin color of whatever body contains him—be it an android body or a flesh body. Either that, or there's something very illogical between issues #438 and #445.

In "The Brainiac Trilogy" (*Action Comics* #649), Brainiac says that Colu's computer tyrants were responsible for disintegrating his real body and that his brain surged across space to enter the skull of Milton Moses Fine. After his Fine body is dunked in recombinant DNA and bionic baths, Brainiac has all green skin, the muscles of a young man in his prime, blond hair, and electronic nodes stuck to his head and back. Somehow, he uses Fine's brain for data storage. In fact, he often complains in various comics that Fine's brain lacks sufficient data storage for Brainiac's great mental agility.

The animated *Superman* series on Cartoon Network tells us that Brainiac is a sophisticated artificial intelligence, a sentient computer system created on Krypton. And there's yet another twist, as seen in the Xbox gaming system's *Superman: The Man of Steel.* In the game, a villain named Brainiac 13 tries to turn Metropolis into a gigantic computer. As Superman battles Brainiac 13, the city is flooded with nanobots and weird robots that are trying to build an enormous Brainiac 13 robot. The game is somewhat consistent with the 1999 *Superman Y2K* series. In *Superman Y2K*, Brainiac pops his mind-essence into a robotic body (exit Milton Fine), calling himself Brainiac 2.5. On New Year's Eve 2000, he tries to download a future version of himself (from God knows where), Brainiac 13. While Brainiac 2.5 enters and takes control of the brain of Lex Luthor's daughter, Brainiac 13 upgrades Metropolis into the futuristic city portrayed by the game.

Well, however the Brainiac story twists, some basic elements remain the same: Brainiac is an artificially intelligent computer built by Colu scientists. He is essentially a form of software. He is evil. Brainiac wants to take over the world.

But how realistic is he? Let's analyze the main science behind Brainiac: artificial intelligence, or AI as it is commonly called. What *is* it? Could an alien computer take control inside the human brain? How similar are the human brain and a computer? Can a computer use the brain for data storage? Let's start with the basics.

By definition, AI has to do with the ability of computers to think independently. Of course, the concept revolves around the basic question of how we define intelligence. Machine intelligence has always been a compromise between what we understand about our own thought processes and what we can program a machine to do.[2]

Norbert Wiener, one of the greatest scientists of the twentieth century, was among the first to note the similarities between human thought and machine operation in the science of cybernetics, of which he is a cofounder. Cybernetics is derived from the Greek word for *helmsman*. Typically, a helmsman steers his ship in a fixed direction: toward a star or a point on land, or along a given compass heading. When waves or wind throw the ship off this heading, the helmsman brings it back on course.

This process, in which deviations result in corrections back to a set point, is called negative feedback. (The opposite, positive feedback, occurs when deviations from a set point result in further deviations. The arms race between the United States and the Soviet Union during the cold war is the classic example.) A common example of negative feedback is a thermostat. It measures and maintains a room's temperature by turning the heat on and off. Wiener theorized that all intelligent behavior could be traced to feedback mechanisms. Since feedback processes could be expressed as algorithms, theoretically intelligence could be built into a machine.

This simple way of looking at human logic and applying it to machines provided a foundation for computer science theory. Early artificial intelligence attempted to reduce our thought processes to purely logical steps, then encode the steps for use by a computer.

A computer functioned at its lowest level by switching between two states: binary one for true and zero for false. Circuits were made from combinations of ones and zeros, thus carrying some inherent limitations. Computers could calculate only through long chains of yes-no, true-false statements of the form "if A is true, go to step B; if A is false, go to step C." Statements had to be entirely true or entirely false. A statement that was 60 percent true was vastly more difficult to deal with. (When Lofti Zadeh, a founder of fuzzy logic, began introducing partially true statements into computer science in the 1970s and 1980s—for example, "The sky is cloudy"—many logicians argued that this was not an allowable subject. The field of logic that deals with partially true statements is called fuzzy logic.)

Ambiguity, error, and partial information were much more difficult to handle. Computers, whose original function, after all, was to compute, were much better equipped to deal with the clean, well-lighted world of mathematical calculation than with the much messier real world. It took some years before computer scientists grasped just how wide the chasm was between these worlds. Moreover, binary logic was best suited to manipulating symbols, which could always be represented as strings of ones and zeros. Geometric and spatial problems were much more difficult. And cases where a symbol could have more than one meaning provoked frequent errors.

Based on yes-no, if-then, true-false statements, the older school of AI is the top-down approach—the heuristic if-then method of applying intelligence to computers. Very methodical. A breakthrough decade for top-down AI was the 1950s. Herbert Simon, who later won a Nobel Prize for economics, and Allen Newell, a physicist and mathematician, designed a top-down program called Logic Theorist. Although the program's outward goal was to produce proofs of logic theorems, its real purpose was to help the researchers figure out how people reach conclusions by making correct guesses. Logic Theorist was a top-down method because it used decision trees, making its way down various branches until arriving at either a correct or an incorrect solution.

Using this approach, Logic Theorist created an original proof of a mathematical theorem, and Simon and Newell were so impressed that they tried to list the program as coauthor of a technical paper. Sadly, the AI didn't land its publishing credential. The journal in question rejected the manuscript.

In 1956, Dartmouth College in New Hampshire hosted a conference that launched AI research. It was organized by John McCarthy, who coined the term *artificial intelligence*. In addition to McCarthy, Simon, Newell, and Logic Theorist (we must list the first recognized AI program as a conference participant), the attendees included Marvin Minsky, who with Dean Edmonds in 1951 had built a neural networking machine from vacuum tubes and B-24 bomber parts. Their machine was called Snarc.

As far back as the 1956 conference, artificial intelligence had two definitions. One was top-down: make a decision in a yes-no, if-then, true-false manner—deduce what's wrong by elimination. The other was quite different, later to be called bottom-up: in addition to yes-no, if-then, true-false thinking, AI should use induction and many of the subtle nuances of human thought.

The main problem with the top-down approach is that it requires an enormous database to store all the possible yes-no facts that a computer would have to consider during deduction. It would take an extremely long time to search that database and to arrive at conclusions. The computer would have to make its way through the mazes upon mazes of logic circuits. This is not at all the way humans think. An astonishing number of thoughts blaze through the human brain simultaneously. In computer lingo, our brains are massive parallel processors.

Top-down AI brings to the table symbolic methods of representing some of our thought processes in machines—that is, top-down AI programs convert known human behaviors and thought patterns into computer symbols and instructions. Perhaps the greatest boost to the top-down philosophy was the defeat of world chess champion Gary Kasparov by the IBM supercomputer Deep Blue. Though not artificially intelligent, Deep Blue used a sophisticated if-then program in a convincing display of machine over human.

Chess, however, is a game with a rigid set of rules. Players have no hidden moves or resources, and every piece is either on a square or not, taken or not, movable in a well-defined way or not. There are no rules governing every situation in the real world, and we almost never have complete information. Humans use common sense, intuition, humor, and a wide range of emotions to arrive at conclusions. Love, passion, greed, anger: How do you code these into if-then statements?

From the very beginning of AI research, there were scientists who questioned the top-down approach. Rather than trying to endow the computer with explicit rules for every conceivable situation, these researchers felt it was more logical to work AI in the other direction—the bottom-up approach. Figure out how to give a computer a foundation of intrinsic capabilities, then let it learn as a child would, on its own, groping its way through the world, making its own connections and conclusions. After all, the human brain is pretty small and doesn't weigh much, and it is not endowed at birth with a massive database having full archives concerning the situations it will face.

Top-down AI uses inflexible rules and massive databases to draw conclusions, to "think." Bottom-up AI learns from what it does, devises its own rules, creates its own data and conclusions—it adapts and grows in knowledge based on the network environment in which it lives.

Rodney Brooks, a computer scientist at Massachusetts Institute of Technology, is one of bottom-up AI's strongest advocates. He believes that AI requires an intellectual springboard similar to animal evolution—that is, an artificially intelligent creature must first learn to survive and prosper in its environment before it can tackle such things as reasoning, intuition, and common sense. It took billions of years for microbes to evolve into vertebrates. It took hundreds of millions of years to move from early vertebrates to modern birds and mammals. It took only a few hundred thousand years for humans to evolve to their present condition. So the argument goes: the foundation takes forever, yet human reasoning and abstract thought take a flash of time.[3]

Therefore, current research emphasizes survival skills such as robotic mobility and vision. Robots must have visual sensors and rudimentary intelligence to avoid obstacles and to lift and sort objects.

Based on the definitions of top-down and bottom-up AI, it makes sense to conclude that *Brainiac is bottom-up AI software*. He most definitely adapts to changes in his environment. When his fellow evil computer tyrants die, Brainiac escapes from Colu, travels through outer space, and lands on Earth. Having lost his original body, Brainiac penetrates the skull of Milton Moses Fine and takes over the human's brain.

Having determined that Brainiac is a bottom-up AI, let's ponder whether Brainiac can indeed control Fine's flesh-based brain: Just how similar is the human brain to a computer? Would it really be so simple to run a computer software program using the human brain as the hardware?

There are some very broad similarities between brain and computer:

- Inputs
- Outputs
- Extreme complexity
- Physical components (brain tissue/hardware)
- Nonphysical components (mind/software)

The brain and the computer have some obvious things in common. Yet despite their similarities, the two are very different.

The basic circuitry in computers relies on the on-off, true-false popping of microswitches. Neurons in our brain also have on-off, true-false states: excited and inhibited. When the voltage across a membrane rises sharply, the neuron is excited and releases chemicals (neurotransmitters) that latch onto receptors of other neurons. When the voltage drops sharply, the neuron is inhibited. Seems awfully similar to the binary on-off states of the computer, doesn't it?

But if we look more closely at neural processes, we see a huge difference. Neurons actually behave in an analog rather than a digital manner.[4] Events leading to neural excitement build up, as if climbing a hill—this is a feature of analog signals. In addition, ions may cross the cell membrane even if neurotransmitters aren't received, and these ions may excite the neuron anyway. Sometimes, a neuron oscillates between intense and minor excitement levels without any outside stimulation. The more a neuron excites itself, the more prone it will be to outside stimulation.

There are approximately fifty shapes that can change the state of the neuron from excited to inhibited, or vice versa. For example, an incoming signal becomes weaker as it traverses a really long dendrite to the neuron body. A signal that travels along a short dendrite will be much more powerful when it hits the neuron body. In addition, it takes a higher dose of neurotransmitter to excite a large neuron than to excite a small one.

The human brain contains approximately one hundred billion to two hundred billion neurons that fire about ten million billion times per second. Each neuron connects to roughly ten thousand other neurons. This is how the brain handles trillions of operations per second. It's an extremely complex neural network.

A computer neural network is a simplified version of a biological one. In the biological form, a neuron accepts input from its dendrites and supplies output to other neurons through its axons. The neuron applies weights to the connections, or synapses, between dendrites and axons. A higher weight might be applied to a synapse related to touching fire than to a synapse associated with seeing the pretty color of fireball orange.

In the computerized version, each input neuron feeds information into every neuron in the inside layer, which may have one or multiple layers of neurons. If the inside layer has two layers of neurons, for example, then every neuron in the first inside layer feeds into every neuron in the second. Every neuron in the last inside layer feeds into neurons in the output layer.

The designer of a neural network provides different weights for the connections among neurons. While our brain receives input from many sources, such as sensations on our skin, what we hear, what we smell, and so forth, an artificial neural network takes input only from values we provide, then weighs everything and supplies a best-guess answer.

Various methods exist for applying weights to artificial neurons and for assembling the input, inside, and output layers into network architectures. A very common neural net architecture is called backpropagation, which compares forecasts to actuals, then adjusts the weighted interconnections among neurons. Over time, as it compares more forecasts to actuals, the neural weights become more accurate. In a sense, the neural network itself has learned and adjusted to its environment.

Nanotechnology guru Eric Drexler believes that nanotech is required to achieve true artificial intelligence in our computer equipment, both hardware and software. He writes,

> One can imagine AI hardware built to imitate a brain not only in function, but in structure. This might result from a neural-simulation approach, or from the evolution of AI programs to run on hardware with a brainlike style of organization. . . . [Speed and accuracy estimates are] crude, of course. A neural synapse is more complex than

a switch; it can change its response to signals by changing its structure. Over time, synapses even form and disappear. These changes in the fibers and connections of the brain embody the longterm mental changes we call learning. . . . Just as the molecular machinery of a synapse responds to patterns of neural activity by modifying the synapse's structure, so the nanocomputers will respond to patterns of activity by directing the nanomachinery to modify the switch's structure. With the right programming, and with communication among the nanocomputers to simulate chemical signals, such a device should behave almost exactly like a brain.[5]

Already, we're wondering how Brainiac, who is bottom-up AI software, can run on (that is, operate on or inhabit/use) the neural circuitry of Fine's human-flesh brain. On the distant planet Colu, did the computer scientists know the structure and complex functioning of the human brain, as found on faraway Earth? If so, why would they pattern Colu AI on Earth brains rather than on Colu brains? We are told that they based Brainiac on the mental patterns of Colu scientists. It seems highly unlikely that the mental patterns of Colu scientists are the same as the mental patterns of Earthlings.

Possibly, Colu scientists built Brainiac's AI using some sort of molecular hardware backbone with nanotech components. If indeed they avoided using binary electronics, then perhaps Brainiac could run on molecular fleshware, such as the neural networks in a human brain.

Luckily for Brainiac, even on primitive Earth—well, we are scientifically primitive compared to Colu—research is already underway to create molecular-based computers. In fact, Stanley Williams, director of the quantum structures research initiative at Hewlett-Packard Labs, claims, "The age of computing hasn't even begun yet. We're still playing around in essentially Stone Age times technologically."[6]

For example, DNA computers have been in the works for years. In 1997, two researchers at the University of Rochester, Animesh Ray and Mitsu Ogihara, constructed logic gates using DNA molecules, a major step toward developing DNA computers capable of solving problems normally handled by digital computers. Instead of using silicon chips and electrical currents, DNA computers rely on deoxyribonucleic acids as memory units and carry out fundamental operations by recombinant techniques. The main difference between DNA computers and electronic computers is that regular computer bits have two positions (on/off) whereas DNA bits have four (adenine, cytosine, guanine, and thymine). Therefore, DNA molecules can in theory handle any problem done on a conventional computer but can also manage more complex operations by using their extra two positions.

Most electronic computers handle operations linearly—one operation at a time, though at incredible speeds. DNA computers rely on biochemical reactions that work in parallel. A single operation in a DNA computer can affect trillions of other DNA strands. DNA computers are thus much faster than electronic computers.

Synthesized DNA strands are used in DNA computers. The amount of information that can be stored in these biological strands is staggering. One cubic centimeter of DNA material can hold up to 1,021 bits of information. More to the point, it's estimated that one pound of synthetic DNA has the capacity to store more information than all the electronic computers in use in the world today.

We conclude that *if Brainiac is a bottom-up AI who was created to run on molecular-flesh hardware—specifically, the neural networks of the human brain—then he is somewhat plausible.* One fundamental question remains: Can the human brain be used for storage by Brainiac's computer mind? Frankly, we find it hard to believe. Memory in humans is not as simple as data storage in computers. In fact, the two are quite different.

In a human brain, memory is spread everywhere; it is not composed of discrete units that are arranged in binary trees or sequential arrays for data searches and retrievals. Rather, many parts of the brain participate in the storage of memory, including the hippocampus, medial thalamus, and temporal lobe. The hippocampus and medial thalamus communicate with the prefrontal cortex, which coordinates the memory of facts with our recollection of when things happened and in what context they happened. This way, we can remember one set of facts in multiple contexts. Someone might recall, for example, that he played basketball with a friend last week and that the friend shot the ball in a particular way. This person might also remember that he played basketball with the same friend two weeks ago and that the friend shot the ball the same way. In fact, our basketball player might remember the same shot from the same angle a hundred times or more.

If Brainiac is built on human brain–like neural networking, then it's conceivable that Brainiac also knows how to use the human brain's extremely complex method of memory storage and retrieval. But there is really no scientific way to stretch ourselves to the point where we can describe how Brainiac accomplishes this amazing feat.

Remember Milton Moses Fine's brain tumor? The tumor is trying to oust Brainiac from Fine's head, but Fine is well on the way to doing so himself. As an alcoholic, he is killing not only his own brain cells but by extension the very cells that contain Brainiac. So while technology sometimes bites back, it seems we can sometimes bite back at technology.

4

Feathers and Fury

The Vulture

Not all superheroes were created equal. Nor were the villains they fought. The Fantastic Four's first foe was the Mole Man and his legion of underground troglodytes. The Incredible Hulk battled the U.S. Army and the Russian fiend, the Gargoyle. Powerful, tough enemies, they tested the new Marvel wave of superheroes to the limit. And then there was Spider-Man.

In his first adventure, published in *Amazing Fantasy* #15, in the guise of Spider-Man, Peter Parker fought a sideshow wrestler called the Crusher. Parker then captured the petty crook who had shot and killed his Uncle Ben. Neither foe proved much of a challenge for the teenager with the powers of a human spider.

When Spider-Man appeared in the first issue of his own comic book, his initial foes weren't really enemies. Instead, they were the Fantastic Four. Being new to the superhero business and the sole support for his aunt and himself, Peter Parker decided the best way to make money was to join the Fantastic Four. The top superhero team in the city was living in a penthouse on top of the Baxter Building, so Peter reasoned that they were making big bucks. After fighting the team to a standstill, Parker was disappointed to learn that the Fantastic Four was a nonprofit group that donated all money and rewards they earned to charity.

In the same issue, Peter found himself battling for his reputation when confronted by a master spy called the Chameleon. Though possessing no real superpower, the Chameleon proved quite a challenge with his quick-change antics, disguising himself as Spider-Man to throw the police off his trail. At the end of the adventure, a despondent Peter walked away from the scene of the crime wondering if he would ever become a true superhero. All of which provided the perfect lead-in to the main story in *Spider-Man* #2, which introduced the first supervillain in the Spider-Man saga, the Vulture.

In "Duel to the Death with the Vulture," the story opens with the Vulture dropping from the sky and stealing a briefcase filled with valuable bonds. From there, the story shifts to Peter Parker in high school worrying about how he and his frail Aunt May will pay the mortgage now that Uncle Ben is dead. Thought balloons, a device used sparingly in comics up to that time, fill every blank space in the panel, revealing the focal character's every thought and emotion. This is done for the main villain as well as the hero.

In their first face-to-face meeting, the Vulture catches Spider-Man by surprise and tosses him into a half-filled water tank. In escaping the trap, Spider-Man is forced to use his brains as well as his brawn. Readers understand exactly how Parker figures the way out of his predicament thanks to the thought balloons.

Although they were a cumbersome device, thought balloons worked much better than the usual comic book gimmicks. No longer did the hero have to explain later in the story how he managed to escape a death trap to a concerned sidekick (a popular *Batman* device). Nor did he need to brag about his triumph to the captured and immobilized crook (as often happened in *Superman* and other DC Comics of the period). Spider-Man grew in popularity because his problems and their solutions were spelled out in black and white. Needless to say, thought balloons soon became standard for every hero and villain in the Marvel universe.

Despite being the first supervillain featured in the Spider-Man saga, the Vulture was a fairly unimpressive foe. He didn't present much of a challenge, even for a young Peter Parker just learning how to use his powers (and discovering such important crime-fighting lessons as never let yourself run out of web shooter fluid). A large portion of the second issue of *Spider-Man* concentrated on the mundane aspects of the ongoing Parker saga. Readers wondered if Aunt May might lose the house that Uncle Ben had bought, if Peter Parker would ever find a part-time job to help pay the bills, and if there was any way Peter could convince the gang at school that he wasn't a geek. These melodramatic cuts relegated the menace of the Vulture to a secondary spot in the issue. As was often the case in subsequent Marvel Comics, *Spider-Man* was soap opera with supervillains.

In his first comic appearance, the Vulture wasn't given a name or any background. He was merely a smart crook looking to make a few big scores. He was an older man, completely bald, with narrow, thin features and a hook nose, thus giving him the appearance of a giant vulture. He wore a green costume made out of synthetic stretch fabric that covered him from neck to toe. Gigantic synthetic feathers on his arms served as his wings. He also had tail feathers connected to his lower spine. A white fur collar around his neck supposedly made him look even more like a vulture.

The Vulture had no visible exoskeleton, yet his legs remained straight behind him when flying. Much like real-life vultures, he made no noise when he flew. However, the lack of sound was not from gliding on air currents like his namesake but was the result of his using an electromagnetic harness that he had invented to supply him with power for his wings. His hideout was an abandoned silo on Staten Island just minutes from New York City. In real life, the Vulture could have made millions legally selling copies of his flying costume to frustrated commuters from Staten Island and New Jersey heading into Manhattan in the morning.

It wasn't until after several encounters with Spider-Man that we learned the Vulture had once lived a normal life before becoming a criminal mastermind. Other than prison fatigues, the Vulture never seemed to wear anything but his green stretch outfit, which maybe explained his lack of a social life. The Vulture's real name was Adrian Toomes, and he had been a brilliant electronic engineer before turning to a life of crime. Toomes and various doppelgangers fought

Spider-Man a number of times, until finally the Vulture joined the Sinister Six, a group of Spider-Man enemies dedicated to combining their powers to defeat their common enemy. Minor villains in the Marvel universe believed in strength in numbers and were always combining into teams like the Sinister Six, the Frightful Four, and the Mutant Liberation Front. Considering their lack of success, these teams might have functioned better as social clubs.

Though the Vulture was described as using an "electromagnetic graviton harness"[1] to increase his strength and stamina when flying, exactly how the device worked was never mentioned in the comic. In several of their aerial duels, Spider-Man used a homemade gadget to scramble the electromagnetic waves generated by the harness, causing the Vulture to plunge earthward.

It was implied in all of his appearances that the Vulture often used his artificial wings as gliders and thus was able to fly noiselessly through the brick canyons of New York City. For a man gliding on wings made out of gigantic imitation feathers, the Vulture performed a number of incredible aerial stunts and vertical ascents that contradicted the laws of physics, but no one seemed to care. Still, the graviton generator was not enough to keep the Vulture airborne. In a decisive fight high above the city, Spider-Man used his webbing to tie the Vulture's wings tightly together and the bird-man dropped like a rock. Only the combination of wings and generator allowed the Vulture to fly.

Can an ordinary, middle-aged, bald guy actually soar through the air using a giant set of wings? If not, then why not?

Humans have dreamed of flying for thousands of years. Long before the rise of modern civilization, Greek mythology told the story of the great inventor, Daedalus, and his son, Icarus, taken prisoner by King Minos of Crete and imprisoned in the Minotaur's labyrinth. Determined to escape, Daedalus designed artificial wings for himself and his son, using wax to attach feathers to their bodies. The two escaped Crete, flying to their freedom. Unfortunately, Icarus was so entranced by flying that he flew too high and the heat of the sun melted the wax on his wings. He fell to his death. Daedalus flew to safety but never used the wings again.

Icarus's death meant different things to different people. Those who believed that if the gods wanted humans to fly, they would have been born with wings felt the story was clear vindication of their attitude. Those who saw the event as a noble attempt doomed by a brash young man considered the motto of the story to be "use better wax." As the battle of philosophies raged over the centuries, brave men continued to try to duplicate the action of birds. It was a challenge that baffled great minds for thousands of years. Even Leonardo da Vinci drew up plans for a pair of artificial wings but never actually constructed them.

As science emerged from superstition, scientists studied and analyzed the movements of birds to understand how they flew. Despite learning how wings worked and how birds lifted themselves off the ground, humans found themselves physically unable to duplicate those actions. They were not meant to fly. At least, not using their own muscle power.

Flying using science, not muscle, became a reality in 1783 in France when the Montgolfier brothers invented the first hot air balloon. A large bag was constructed from paper and linen, and a hot fire was positioned on a platform attached beneath it. The heat generated by the fire caused the air inside the bag to expand, squeezing air molecules from the bottom opening. Thus, the air inside the balloon was lighter than the air outside. The lighter-than-air balloon flew upward, carrying a goat, a chicken, and a mouse as its first passengers.

A few months later, two brave Frenchmen flew by balloon across the English Channel. The age of lighter-than-air flight had begun. For the next hundred years, scientists and inventors experimented with balloons. More than a few of these experiments ended in disaster, but they didn't discourage balloonists. For long trips, a mixture of lighter-than-air gas (preferably helium instead of the explosive hydrogen) and hot air worked best, and intrepid flyers traveled across America and Europe.

The world changed once more on December 17, 1903, when the Wright brothers flew the first heavier-than-air vehicle, an airplane, at Kitty Hawk, North Carolina. The modern age of flight began. A century of airplanes, large and small, followed. But humans still needed machinery to fly. Why?

The most obvious answer to why the Vulture and all other winged supervillains and superheroes (such as DC Comics' Hawkman and Hawkgirl) can't fly is that they're not birds. While humans and birds are biologically similar in many ways, including having a heart, lungs, and stomach, they are also quite different.

Birds are designed for flying. They have extremely strong hearts and chest muscles. They are lighter than other animals, as their bones are hollow. More astonishing, the skull bones of birds have air cavities continuous with the nasal cavities. Trunk bones like the breastbone, vertebrae, and pelvic bones also contain air sacs. These hollow bones are known as pneumatic bones. Along with making the bird skeleton lighter, they also serve as a source for extra oxygen to be absorbed into the blood for greater energy.

The strong but lightweight bones in the wings and legs of birds are long, hollow tubes supported by many small cross-braces. Also, several smaller bones in birds are fused together, creating one large, strong bone. The bones of the collarbone are connected to provide strong support for the powerful shoulder muscles that move the bird's wings.

Birds do not have teeth or heavy jaws. They use bits of gravel inside their body to grind their food into small pieces. A bird's skull only has two thin layers of bone. Birds also have extremely powerful lungs that efficiently remove oxygen from the air and filter it into their blood. All of these factors, combined with the shape of their wings and the composition of their feathers, make it possible for birds to fly. Genetic engineering won't be producing any bird-men in the near future. For the record, the heaviest bird capable of flight is the great bustard, which weighs approximately forty pounds. Since our criminal mastermind, the Vulture, is a lot heavier, he's not operating under bird power. Which means he's flying like a human airplane.

• • •

To truly understand how birds and airplanes fly, we must define the four basic terms of aerodynamics. *Thrust* is the force generated by the plane or bird to move forward. Airplanes in the twenty-first century normally create thrust by using jet engines. Birds create thrust by flapping their wings. The aerodynamic force overcome by thrust is known as *drag*.

The third force acting on a flying object is its *weight*. One fact worth remembering is that everything on Earth, including air, has weight. Obviously, the weight of a plane or a bird is what keeps it on the ground. The aerodynamic force that raises and holds the airplane or bird in the air is known as *lift*. When lift is greater than weight, a plane flies. When lift decreases, the plane descends. Both lift and drag can only exist in a moving fluid—and air, in this case, is considered to be a fluid. Thus, neither birds nor airplanes can work in outer space where there is no fluid.

Of these four terms, the first three are easy to understand. A motor powers an airplane and muscles flap a bird's wings, in each case developing thrust. Air resistance acts as drag. The weight of an object is its mass affected by the power of gravity. Only lift is a mystery. How is lift created?

There are two fairly simple scientific methods used to explain lift in most textbooks: Bernoulli's principle and the transfer of momentum principle. Unfortunately, while each of these standard explanations is partially right, each is also partially wrong. We'll describe the two wrong explanations so that you can recognize them (and be a hero to your class by correcting your science teacher), understand their flaws, then examine the truth about lift and what it tells us about airplanes, birds, and a certain crook called the Vulture.

The Bernoulli principle is often called the longer path explanation. It looks at the top and bottom surfaces of an airplane wing, and deals with a stream of air particles traveling toward the front of the wing. When the particles split at the wing tip, the ones traveling over the top have a greater distance to travel to the back. If they are going to take the same amount of time to make it to the back of the wing as the particles going underneath, they must travel faster. Bernoulli's equation, one of the fundamental rules of fluid dynamics, says that as the speed of a fluid increases, the pressure it exerts decreases. Thus, Bernoulli's equation implies that the pressure on top of the wing must be less than the pressure on the bottom of the wing. Since the air pressure beneath the wing is greater than that above the wing, the wing (and with it, the airplane) rises.

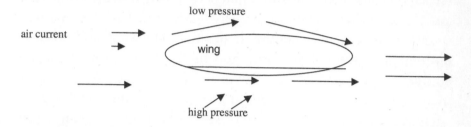

Unfortunately, there are several major problems associated with applying Bernoulli's principle to wings. For one, there's no valid reason why the air particles that go over the wing and under the wing need to meet at the back of the wing at the same time. Another problem is that not all wings have a curve on top and a flat surface on the bottom. Some wings are curved both above and below. More troublesome, sometimes planes fly upside down. If Bernoulli's principle were true all the time, the higher air pressure would be pressing down on the top of the wing and would drive the plane straight into the ground.

Still, the longer path explanation isn't entirely wrong. The air on the top part of the wing does flow faster than on the bottom. So there is some truth to the theory.

The transfer of momentum theory is based on Newton's third law: for every action there is an equal and opposite reaction. Newton imagined air molecules acting like bullets and striking the bottom surface of the wing. These particles would add the force of momentum to the wing and slowly move it up into the air.

The problem with Newton's idea is that air acts as a fluid, not as a stream of molecules. Also, his theory never takes into account the top surface of the wing, so his calculations are not very accurate. However, at very high speeds (five times that of sound), air molecules behave much more like bullets, so Newton's ideas aren't entirely worthless.

If we take some of the best of both theories, we finally come up with an explanation that has no flaws and explains lift clearly and concisely. Lift is a force on a wing completely immersed in a moving fluid (air). It acts on the wing in a direction perpendicular to the flow of the air. The force is created by differences in pressure that occur because of variations in the speed of the air all around the wing (both top and bottom). The result of this force is divided into lift (raising the wing) and drag (slowing it down). When the flow of air past a wing is increased, the pressure differences between the top and bottom of the wing become greater, and lift increases. Lift can also be changed by varying the angle of the wing.

Putting all these factors together, we come up with the standard equation used for calculating lift:

$$L = (C)(R)(V^2)(A)$$

where L = lift
C = the lift coefficient
R = air density
V = air velocity
A = wing area

Looking at the equation, we immediately notice that lift is dependent on two variables: the air velocity (or how fast the plane or bird is traveling) and the wing area. For people who are deathly afraid of flying, it's worth noting that a 747 generates a lift greater than 870,000 pounds on takeoff. The lift coefficient is entirely dependent on the angle of the wing, while the air density is totally dependent on the height of the plane or bird above sea level.

Finally, we can plug in some numbers. If we are flying at sea level, R = .00237 slugs/cubic foot (taken from a handy textbook defining air density). Using a table of lift coefficients calculated by the National

Advisory Committee of Aeronautics for a 1408 airfoil shape with the wing at 4 degrees,[2] we arrive at $C = .55$. Doing some quick calculations, $L = .00065175 (V^2)(A)$. Which again points out that the lift of a plane flying at a certain height (in this case, sea level) with a wing at a specific standard angle is directly related to the air velocity and the area of the wing. So, the weight of our plane or bird (or bird-man) relies on how fast it moves and the size of its wings.

Now the numbers get interesting. If the Vulture weighs 200 pounds (fully equipped with wings and costume) and is sprinting at 20 feet per second (a record-setting pace, running a mile in less than four minutes), he will need a wing area of approximately 800 square feet to lift him off the ground. Of course, that assumes he can keep moving at 20 feet per second all the time he is in the air. Which leads us to conclude that the Vulture flying under his own power or even with the aid of an electromagnetic booster is impossible. Still, he's not as outrageous now as he seemed nearly forty years ago.

In 1933, a German group of aviation experts offered a 5,000-mark prize for the first human-powered airplane. Prizes were offered in Italy and Russia, as well, but the money went unclaimed. The problem, as seen in the equation above, is one of power. Even with huge ultralight wings, it takes at least 10 horsepower to power a glider, and the best-trained athletes can only generate 0.4 horsepower for any length of time.

In 1959, British industrialist Henry Kremer offered 50,000 pounds for a human-powered aircraft that could fly around two markers four-fifths of a mile apart. On August 23, 1977, eighteen years after the prize was offered, California energy consultant Paul McReady claimed the money with his self-designed plane, the *Gossamer Condor*. Later, in the *Gossamer Albatross*, McReady flew across the English Channel.

The *Condor* was the result of years of designing and redesigning a plane aimed solely at winning the Kremer prize. Working with Dr. Peter Lissaman, McReady modified the plane after each test flight, relying on engineering know-how instead of computer modeling.

The plane was made of extremely thin aluminum tubes covered with Mylar plastic and braced with stainless steel wires. The pilot sat in a semireclining position, which enabled him to have both hands free for the controls. One hand controlled the vertical and lateral movement, and the other hand moved a lever controlling steel wires that twisted the wing, making the plane turn.

The pilot who flew the plane was Bryan Allen, a champion bicyclist and hang-glider pilot. The plane reached a speed of approximately 11 miles per hour, with Allen developing one-third horsepower by bicycle pedaling. The *Condor* weighed 70 pounds without a pilot and approximately 200 pounds with Allen. It was 30 feet long, 18 feet high, with a wingspan of 96 feet.

The *Albatross*, an improved version of The *Condor*, weighed only 60 pounds. The energy needed to attain proper velocity was obtained by bicycle pedaling at 20 miles per hour.

Not everyone is willing to fly by the rules. Take, for example, Felix Baumgartner.

On July 31, 2003, the Austrian skydiver became the first person to skydive across the English Channel. Felix jumped out of a plane above Dover, England, and landed just 14 minutes later in Cap Blanc-Nez near Calais, France, 22 miles away. He wore only an aerodynamic jumpsuit with a 6-foot carbon fin strapped to his back, an oxygen tank from which to breathe, and a parachute to land. Baumgartner leapt from the plane when it was 30,000 feet in the air, free falling most of the time at approximately 135 miles an hour.

Human-propelled flying is still in its early stages. Hundreds of young scientists and engineers from around the world are constantly working on more efficient and durable human-powered planes. It's doubtful that they'll ever come up with a green-feathered suit used for criminal activities, but you never can tell. While the Vulture remains strictly in the world of comics for the present, who knows what will happen in the real world during the next twenty years?

5

The Kiss of Death
Poison Ivy

Created by DC Comics and featured in *Batman*, supervillainess Poison Ivy is a potent femme fatale. Her face and hair are gorgeous, and her body is beautiful. But her body is also an instrument of death and mayhem: it manufactures killer toxins, keeps her skin lush with chlorophyll, fills her lips with venom, and pumps so many phero-mones into the air that men fall instantly in love with her. If that weren't enough, her hobbies include creating killer plants. And for those poor men who fall in love with her, did we mention that her kiss spells death?

While Poison Ivy is unique in comics, human/plant hybrids have a long and varied history in science fiction. Murray Leinster's 1935 short story, "Proxima Centauri," features evil tree men as villains who hijack a starship from Earth. David H. Keller's "The Ivy War" from an early issue of *Amazing Stories* describes a mutated ivy plant that takes over a city. In "Plants Must Kill," Frank Belknap Long's space detective, John Carstairs, battles an intelligent plant that mur-ders businessmen in an asteroid city. In Jack Vance's "The Houses of Iszm," the houses are gigantic, intelligent plants that grow rooms to order. In James Patrick Kelly's "Mr. Boy," a house made out of superplant life also serves as a kid's mother. Perhaps the most famous intelligent plant of all (and a possible inspiration for Poison Ivy) is Audry Jr. from the 1960 B-film (and later musicals) *Little Shop of Horrors*.

Poison Ivy makes her first appearance in "Beware of Poison Ivy!" (*Batman* #181, June 1966). At the opening of a special exhibit at the Gotham City Museum, she unleashes a lipstick that detonates all of the newsmen's flash bulbs. Everyone is struck blind, as Batman points out, "Your lipstick sent electrical impulses that exploded the flash bulbs." Later, she kisses Batman's cheek with chloroform-based lipstick and sends him into a dizzy faint, hot with love for her. He is incapable of putting such a "beautiful doll behind bars."

So who is this beautiful doll, and where did she come from? Once upon a time, Poison Ivy was a mild-mannered, mousy botanist named Pamela Isley. Big glasses, snarled brown hair, dowdy clothing. Pamela spent all her time in the lab working on odd plant hybrid projects with the insane Dr. Jason Woodrue. Sadly, Pamela turned into an odd plant hybrid project herself, as Woodrue mutated her body chemistry to be more plantlike. Other versions of Poison Ivy's origin in film and on television suggest that a freakish lab accident with toxic plant matter mutated her body.

Whatever the case, Pamela emerged as a towering goddess: shapely curves, skin-tight green outfits, long, bright red hair, gaudy makeup, and the ability to seduce men with her mere presence. She emerged as a plant-woman who could create killer poisons, man-eating Venus flytraps, and enzymes that turn humans into trees.

A plant-crazed nature lover, Poison Ivy is often called Mother Nature. She protects parks from chemical factories and pollution-pumping corporations. For example, in the animated *Batman* episode "Pretty Poison," she kills District Attorney Harvey Dent with her toxic lipstick because he destroys an endangered plant. The lipstick contains poison from the extinct wild thorny rose.

While her body emits plant-based perfumes and skin toxins that kill other people, Poison Ivy remains immune to her own deadly devices. In fact, nothing seems to kill her.

What is *real* poison ivy? Is it anything like Batman's supervillainess with her killer lipstick? In fact, is Batman's Poison Ivy a plant or a human? And while we're pondering these questions, let's figure out whether Poison Ivy's weapons—pheromones, skin toxins, lip venom, man-eating Venus flytraps—are at all possible.

Real poison ivy is known scientifically as *Toxicodendron radicans*. According to the American Botanical Council,[1] Captain John Smith was the first person in North America to describe and draw poison ivy. He did this in the early 1600s and was also responsible for naming the plant. Today, poison ivy is found in many countries, among them, the United States, Canada, Mexico, England, Malaysia, China, and Japan.

Boy Scouts and Girl Scouts can tell you that poison ivy is a shrub or ropelike vine with three green leaflets per main stem, with the middle one protruding from a slender stem connected to the main stem. In late spring, the leaves produce greenish white flowers. In summer and fall, the female plants produce white berries. When cut, poison ivy stems exude a sticky black sap. This sap is also present in tiny resin canals within the roots, leaves, flowers, and berries.

Also within the resin canals is the allergen known as urushiol, which is a mixture of phenolic compounds called catechols. These phenolic compounds (pentadecylcatechols) are comprised of a benzene ring with fifteen carbon side chains:

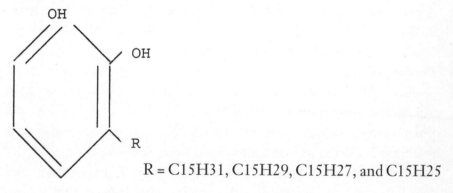

$$R = C15H31, C15H29, C15H27, \text{ and } C15H25$$

If stems, leaves, flowers, or berries are bruised in any way, then urushiol may be on the surface of the plant as well as in the resin canals. Smoke from burning poison ivy plants, as well as forest ash and dust, may contain urushiol. Reactions to urushiol include redness, swelling, itchiness, and blisters. In fact, 80–90 percent of adults get rashes if exposed to 50 micrograms of urushiol.[2] To give you an idea of the tiny amount of urushiol required to trigger a poisin ivy rash, a single grain of salt weighs 60 micrograms. So if poison ivy sap that is the size of one grain of salt happens to get on your skin, you may have an allergic reaction in the form of a rash that can spread all over your body. According to some experts, the amount of urushiol on a pin-head can trigger allergic reactions in five hundred people.[3]

According to the Food and Drug Administration (FDA),[4] urushiol sticks to "pets, garden tools, balls, or anything it comes in contact with." Indeed, its potency lasts for many years, sometimes decades, so your dog, rake, or basketball could have contact with poison ivy and you could get the rash years later. Just touching something that has urushiol on it causes the allergic reaction.

It is almost impossible to wash urushiol off your skin. If you scratch it, the rash spreads. As soon as urushiol hits you, it penetrates the epidermal or outer layer of your skin, binding to your skin cell membranes and sometimes entering your bloodstream. Some researchers claim that if urushiol enters your bloodstream, you'll have new eruptons of poison ivy every few hours.[5]

Once attached to your cells, urushiol attracts the effector T cells of your body's immune system. Eventually, these cells release lymphokines, which attract killer T cells. These killer T cells release toxins that destroy urushiol. When your body attacks urushiol and all the skin cells bound to it, that's when you start itching like crazy.

As of this writing, the FDA has not approved any vaccines or creams for protection against poison ivy.[6] Of course, creams and other ointments do help relieve itching. Corticosteroids, which are hormones, and manganese sulfate solution are often recommended.[7] But a cure: not yet. Though not approved by the FDA, prescription pills may offer immunization from poison ivy—at least, that's what some researchers believe.[8] The medication must be taken for months and may cause "uncomfortable side effects."

Because even dead poison ivy causes allergic reactions, it is extremely difficult to eliminate the danger of the plant. You can try using herbicides, but because poison ivy consists of a vast network of rhizomes beneath the ground, they rarely destroy the plant. But herbicides will kill the surrounding plants.

Yet when Poison Ivy, the villainess, kisses men, they do not break out in hideous rashes; nor do they scratch. They die. Apparently, her body does not exude urushiol. So what could cause this effect?

Just as molds and fungi can grow on human skin, causing, for example, athlete's foot, perhaps Poison Ivy's lips are coated with a plant substance. In her case, the plant substance not only produces the usual chloroplasts and chlorophyll, but perhaps the DNA within the plant cells is mutated to produce human toxins. This would explain why her kiss is deadly, with or without lipstick. In fact, the fungus need not mutate. If Poison Ivy is coated with a common mushroom toxin, then kissing her would kill you. Poisonous mushrooms are aplenty, as are their toxins: amanitin, gyromitrin, orellanine, muscarine, ibotenic acid, muscimol, psilocybin, and coprine are all produced by mushrooms.

If anything, it's surprising that Poison Ivy doesn't have more than a deadly kiss. If the poisonous plant—be it a mushroom or something else—grows on her lips, why doesn't it grow elsewhere on her body? Or does it?

Perhaps the skin-tight green leotards hide a deadly toxic wasteland that grows from her breasts to her hips. Perhaps beneath that wild mane of red hair is a deadly trap of mutated mold and mildew.

As for attracting men in the first place, Poison Ivy only needs a little perfume—that is, naturally occuring pheromones, which are chemical substances that animals secrete to convey information and to excite other animals of their species.

Many species of beetles, bees, and moths make great use of sex-attractant pheromones. In fact, the polyphemus moth only mates if the female secretes male-attracting pheromones, which she only does if she's wandering through a pile of red oak leaves. The leaves emit a volatile aldehyde that stimulates her production of mating pheromones.[9]

The study of mammal pheromones is still evolving. Much work has been done regarding sensory neurons from the vomeronasal organ (VNO), a pheromone-sensing structure in the nasal cavities of animals such as dogs, cats, elephants, and mice. Because the VNO is in the nasal cavity, you might think that the pheromone-sensing structure is wired into the olfactory system. However, this is not the case. The VNO plays a far more important role than simply identifying perfumes, musks, and other odors. It actually controls gender recognition.[10]

It is quite likely, therefore, that Poison Ivy secretes pheromones that attract males who find her irresistible. In fact, using her acute pheromone system, Poison Ivy could also affect men in many other ways. For example, Catherine Dulac of Harvard reports[11] that the VNO pheromone system controls genetically preprogrammed territorial, social ranking, and maternal behaviors.

This means Poison Ivy's pheromones could cause calamities such as Batman becoming highly aggressive and territorial, attacking anyone and everyone who vies with him for control of Gotham City; mothers all over Gotham City leaving their babies and children homeless in the gutters; mayors, governors, and even the president of the United States and foreign leaders resigning from their posts and giving their positions to janitors, tree debarkers, and fiction writers.

In addition to killer pheromones, Poison Ivy battles men with killer lipsticks in all the latest shades. Her lipstick comes in a chloroform base, and it explodes flash bulbs, causing men to go blind. If that weren't enough, a brush of her lips, a mere touch of a kiss, kills people.

How possible is chloroform-based lipstick? Can lipstick explode flash bulbs, causing temporary blindness? Can your touch kill? And what about Poison Ivy's lip venom: is she a viper, or not?

Let's analyze that lipstick. Over-the-counter lipsticks consist of a base, such as oil and wax, plus flavors, colorants, and perfumes. But lipsticks are easy to make at home, too; or in an insane botanist's evil lab. For example, anyone can make lip balm from grated beeswax, oil, and a dash of mint flavor. This all sounds pretty simple. Now let's determine whether you—or an insane botanist—can add chloroform to homemade lipstick.

First, some definitions. Chloroform ($CHCl_3$) is a colorless liquid with a sweet but burning taste. In heat, chloroform yields toxic fumes made of phosgene and hydrochloric acid. It can burn your skin, give you a sore throat, and cause cancer. It can kill body cells. In 1976, the Federal Drug Administration banned the use of chloroform in cosmetics and drugs.[12]

It's interesting to note that chloroform can indeed be administered via oil. Fat retains unmetabolized chloroform longer than other body tissues. If Poison Ivy makes her lipsticks with oil and beeswax, then it's possible for her to infuse the oil part of the concoction with chloroform.

As for the beeswax in Poison Ivy's lipstick, it dissolves partially in chloroform, as well as in other substances such as ether and carbon tetrachloride. In addition, beeswax mixes well with many oils. So it would seem that even if the chloroform in Poison Ivy's lipstick partially dissolves the beeswax, the combination of beeswax and oil should suffice to contain some amount of chloroform. Probably not enough to kill a man, but then, her kiss usually leaves men staggering, faint, and dizzy.

To make it a kiss of death, she could pull out a special lipstick— say, Poison Ivy Psilocybin Ghastly Green. Not only would this special ("lasts all day") lipstick contain gobs of killer fungi to complement the fungal (or other deadly) growths upon her lips, it would also have a thick oil-and-chloroform component: perhaps this is not a lipstick at all but rather an extra-slick high-sheen lip gloss. How chic!

With all this noxious matter upon her lips, does Poison Ivy still need the venom that fills her lush lips?

Hey, a woman can never be too attractive. If you're going to sport Psilocybin Ghastly Green, you might as well go the extra mile and make sure that when you nibble on the guy's neck, he really falls for you: hard; uh, knocked flat on the ground, in Poison Ivy's case.

Typically, an animal's venom is produced by one or more glands. These glands are connected to a body part that administers the venom to victims. So, for example, snakes and spiders administer venom with their fangs, bees and scorpions use stingers, fish use spikes, centipedes use pincers, millipedes use squirters, and cone shells use poisoned harpoons. The amount of venom varies, and most often it is injected into the subcutaneous layers of skin—that is, the animal does not inject its venom into the victim's internal layer of skin or body organs.

If Poison Ivy's lips are filled with venom, how does she inject it into her victim's lips, mouth, cheek, or neck? We never see fangs on Poison Ivy. Nor do we see stingers, spikes, pincers, squirters, or harpoons coming out of her mouth. Does she perhaps use her teeth?

Lipsticks are only one way Poison Ivy exploits plant technology to her enemy's detriment. While some of her means of causing plant mutations are a bit farfetched, such as using enzymes to turn people into trees, others seem feasible. Take, for example, her creation of the giant Venus flytrap that eats people.

We know of more than six hundred species and subspecies of carnivorous plants. The Venus flytrap (*Dionaea muscipula*) is one of the most famous and gruesome. However, it does not eat animals. While the the tropical pitcher plant (*Nepenthes*) has been known to eat an occasional mouse, frog, or bird, most carnivorous plants confine their prey to insects.

The rare Venus flytrap grows in southern North Carolina and South Carolina. It tends to be green with red splotches on the insides of the traps. Some Venus flytraps are cool and sporty: their traps are splotched with jazzy red, yellow, and green. The traps themselves are bilobed leaves that are lined with hairs. When something touches the hairs, the leaves snap shut.

Just because the Venus flytrap feasts on insects rather than people, this doesn't mean that the plant is not capable of eating humans. In fact, it is possible for an ordinary Venus flytrap to digest human flesh—but only in small quantities.

Were the Venus flytrap twenty times its ordinary size, with enlarged, powerful traps gushing with digestive juices, then it's conceivable that the plant could eat parts of a person. Eventually, if trapped inside a large enough Venus flytrap, a person could be totally consumed. In addition, because the Venus flytrap produces its own digestive enzymes—as opposed to using bacteria to digest its prey—it's possible that a mutated plant could produce especially powerful juices that melt flesh on contact.

With her advanced scientific training and expertise, Poison Ivy could indeed create a mutated plant. Genetic manipulation would do the trick.

Genes determine how plants and animals handle poisons, battle infections and other illnesses, digest foods, and respond to environmental conditions. Genes determine what plants and animals look like.

A genetic mutation may be inherited from one or both parents. When this happens, the mutation is in almost every cell, and when cells divide, the mutation is reproduced in the new cells. This is a germ-line mutation, derived from the notion of germ cells, the egg and the sperm.

Transgenics refers to the creation of embryos containing genes from other species. Specifically,

> Not only can a foreign gene be put into the cells of an organism: the gene can actually be incorporated into the DNA derived from germ cells or embryonic cells of another organism. From this combination, an embryo can be produced that contains this gene that came originally from another species (called a transgene). Transgenic embryos can be put into an adult female . . . which will then give birth to [offspring] permanently carrying the transgene.[13]

Once you accept the concept of transgenics, you can easily imagine its applications. For example, someday we might teach toddlers about new kinds of farm animals. In addition to the traditional cows, lambs, chickens, and pigs kept onsite for that old-time feeling, Old McDonald's Farm might showcase fields of docile pig-lambs, horse-chickens, petunia-cows, and lion-peacocks. It's not as silly as it sounds. Pigs may be bred to have wool coats. Sheep may have bacon-flavored meat. Chickens may shed their feathers for light horse down, and horses may taste like Thanksgiving turkeys. Tuna-textured cows may smell like flowers rather than manure; cows may indeed serve as a source of fish, complete with all the vitamins and nutrients and none of the fat found in traditional beef. Lion manes may look like peacock sprays. And it's also conceivable that female horses may give birth to horse-chickens and petunia-lambs.[14]

Will we have plaid roses and flowers that smell like sea spray and jasmine? Of course, we will! Granted, plaid may be tough to manage. However, blue roses are possible due to transgenics, and certainly fragrance requires minimal tinkering. About the genetic engineering of roses, in particular, Michael Gross, who has a doctorate in physical biochemistry, says the following:

> [Researchers] have isolated thousands of pigments from the petals of different varieties of roses, characterized them, tracked down the enzymes involved in their synthesis, and the physiological conditions required for the proper coloring. After all of this, it dawned on them that blue roses cannot be bred as a matter of principle. All roses known lack the enzyme that would convert the common intermediate dihydrokaempferol to the blue delphinidine-3-glucoside. The only way out of this dilemma is to transfer "blue genes" from different plant species. The DNA sequence encoding the enzyme in petunias could be identified and transferred to petunia mutants whose enzyme was deficient. In principle, it should be possible to transfer the gene into roses as well, and provide them with blue petals.[15]

It's not much of a stretch to imagine Poison Ivy's genetically brewed man-eating Venus flytrap. A little genetic engineering, some transgenics, and voilà: killer plants.

Despite her prowess in the botanical lab, Poison Ivy's supervillain powers lie chiefly in her ability to make men faint: her pheromones are her power. Her venom-filled lips are dangerous, but they aren't armor, and there's nothing much that lips can do against a bullet or even a whack over the head.

If Batman and the cops would wear gas masks and lip guards, it would be simple for them to avoid the power of Poison Ivy's pheromones. Beyond that, a little weed killer could go a long way.

6

Groping for Power
Doctor Octopus

Doctor Otto Octavius, more commonly known as Doctor Octopus or Doc Ock, first appeared in "Spider-Man versus Doctor Octopus" (*Amazing Spider-Man*, vol. 1, #3, 1963). As with many other supervillains, Doctor Octopus proved to be an evil genius with extraordinary powers; in this case, the bone-shattering power of four steel tentacles attached to a harness on his chest and waist.

Presumably, if the harness is removed, then Doctor Octopus no longer has the power of his steel tentacles. But as the story line goes in "Spider-Man versus Doctor Octopus," he cannot remove the harness. In the comic, Doc Ock is a nuclear physicist, and he creates robotic arms with projectile capabilities to help him do research with radioactive substances. He stands behind a lead wall, which shields him from the radiation, and using the four artificial arms as well as his own two hands, he manipulates volatile chemicals and radioactivity. A terrible accident occurs in his laboratory, and Doc Ock is exposed to an enormous amount of radiation that somehow welds the robotic arms to his body. The radiation gives him mental telepathy control over the robotic limbs, simultaneously transforming him into a criminally insane maniac. He decides that he is the most supreme person ever to live. He wants (of course) to destroy Spider-Man and take over the Earth.

In this chapter, we'll question whether radioactivity can graft robotic arms to a human and whether it can trigger insanity. We'll also ponder the notion of the harness being permanently glued to Doc Ock's body. And we'll take a look at advances in prosthetics and bionic limbs. Just how close are we to creating robotic appendages for our bodies? But for now, let's continue with our story.

With the harness on, Doctor Octopus can telepathically control his tentacles to move at 90 feet per second. His brain can tell the steel limbs to strike Spider-Man and other foes with the force of a jackhammer.[1]

And the steel tentacles do more than just clobber people. Using the three pincers at the end of each appendage, Doctor Octopus can poke, prod, throw, and manipulate many objects at once. He also uses his tentacles as a fan: he spins them, producing enormous winds. In addition, he can extend the tentacles to longer lengths so that he can walk and run on them—at up to 50 miles per hour—as if they were stilts. For example, in "Doctor Octopus" (*Marvel Tales* #38), Doc Ock whirls his tentacles madly, blowing away tear gas spewed at him by police. He escapes on his hydraulic arms. Oddly enough, in this tale he does take off the metal torso contraption, and hence, his metal arms. Later, in "Disaster!" (*Marvel Tales* #41), a nullifier machine deactivates his harness, causing the metal arms to fall off his body. This seems in direct contradiction with the origin story, which tells us that the steel limbs are grafted or welded—somehow permanently glued—to his body.

Prosthetic limbs are artificial replacements of flesh-and-blood limbs. They are used by many real (as opposed to comic book) people.

Peg legs are the simplest type of prostheses, and they have no electronic components. Another simple type of artificial appendage is an arm that ends with pincers rather than a hand with fingers. Rather than Doctor Octopus's sophisticated pincer devices, this simple limb is attached to whatever is left of the patient's real arm. It is also attached to a harness that is strapped around the patient's shoulders. When the patient moves his shoulder, the harness moves, pulling cables that open and close the hooks.

However, far more sophisticated devices do exist. Dynamic protheses contain electronic components and are based on myoelectric properties. In short, a myoelectric prosthesis contains sensors that respond to the electricity created by the movement of flesh-and-blood muscles. When a patient tenses his muscle—say, in his upper arm—the sensors in the prosthetic portion of his arm detect the myoelectric transmission and send the corresponding signals to the artificial hand. Run by batteries, the hand opens or closes. Some prosthetic limbs even have sensors that detect temperature. These devices send hot and cold information to electrodes in the skin, enabling a patient to "feel" with his prosthetic limbs.

Today's advances include artificial feet that cushion the body on the ground as if they were real and feet with electronic components that enable patients to balance their weight more evenly. For example, the Elation™ Flex-Foot from a company called OSSUR contains "flex-foot technology" along with adjustable heel heights. The Elation Flex-Foot automatically adjusts its mechanical pieces—known as foot blades and rocker plates—based on the amount of weight placed upon it. If a patient is heavier than average, if he shifts his weight from one foot to the other, or if he leans heavily in one direction, the foot blade presses more strongly against the rocker plate, thus changing the cushioning or impact of the foot against the ground. According to OSSUR, "A narrow, anatomically correct foot cover with a sandal toe contour is bonded to the foot, making it suitable for dress shoes, sandals, cowboy boots and other types of footwear. Elation is easy to cosmetically finish and the foot is ideal for amputees of low and moderate impact levels weighing up to 220 pounds."[2]

For amputees who have lost limbs at the hip level, modern medicine provides artificial hip joints made of laminated plastic or thermoplastic. Prosthetic devices are commonly made from carbon fibers, titanium, and polypropylenes, which are flexible plastics. Prostheses can be constructed of a bulletproof material called Keblar. To make limbs really strong, a prosthetic can be devised of a layer of carbon, a layer of Keblar, and another layer of carbon.

According to *Medical Device & Diagnostic Industry Magazine*,[3] much research is being done to create materials that emulate human muscles. For example, a full-size plastic skeleton named Mr. Boney roams around the University of New Mexico Artificial Muscle Research Institute. Mr. Boney's microprocessor-controlled heart pumps a chemical fluid through his body, and this fluid is what actuates his artificial muscles.

Israeli scientists discovered in the 1940s that polymer fiber gels shrink in acid solutions. They also expand when a base is added to the acid solutions. Apparently, these properties are similar to those of biological muscles. Another way of making "robotic muscles" contract and swell is to expose them to electrical current. Modern research focuses on making artificial muscles respond more quickly and more significantly to chemical and electrical stimulation—that is, we're trying to get artificial muscles to lift more weight and to do it quickly.

Dr. Qi-Ming Zhang, associate professor of electrical engineering at the Pennsylvania State University Materials Research Laboratory, bombards copolymer material with electrons to make it more flexible.[4] That is, as voltage increases, the copolymer material becomes increasingly capable of movement—in fact, it is up to forty times more flexible than other materials used in prosthetics.[5]

It takes approximately 100 milliseconds for human muscles to respond to transmissions from the brain. Using new materials and techniques, modern prosthetics can move at near-human speeds. Unfortunately, the strength of these prosthetics remains limited. This issue has been addressed by Mo Shahinpoor of the University of New Mexico, who has developed artificial muscles made from thousands of strands of polyacrylonitrile. These muscles are twice as strong as human muscles.

By combining artificial intelligence, robotics, sensors, micromachinery, distributed processing, and other technologies, scientists will create a wide variety of smart materials and devices over the next couple of decades.[6] According to a company that specializes in creating them, smart materials are "any material that shows some form of response (often physical) such as mechanical deformation, movement, optical illumination, heat generation, contraction, and expansion in presence of a given stimuli, such as electricity, heat, light, chemicals, pressure, mechanical deformation, exposure to other chemicals or elements. The response may be useful in converting the applied energy into a desired motion or action."[7]

Scientific American says that soon smart devices will be in everything:

> Forget dumb old bricks and mortar: engineers are designing future devices from exotic materials that incorporate chemical switches or mechanical sensors to improve their performance. These "smart materials" are just starting to emerge from the laboratory, but soon you can expect to find them in everything from laptop computers to concrete bridges. Philip R. Troyk of the Illinois Institute of Technology has constructed wireless sensors no larger than a Rice Krispie. Implanted in a patient's muscle, the devices could relay information on local nerve activity via radio to an external computer. The devices could also receive power through magnetic induction and send out mild shocks that stimulate the muscle into action.[8]

Sandia National Laboratories in Albuquerque, New Mexico, is conducting research about embedding these smart materials and devices in walls, doors, and other building structures. Sandia's smart materials will "sense disturbances, process the information and through commands to actuators, accomplish some beneficial reaction such as vibration control." Their work includes flexible robotics, photolithography (the manufacture of smaller microelectronic circuits), biomechanical and biomedical products (artificial muscles, drug delivery systems), and process control (solar reflectors, aerodynamic surfaces).[9]

When we start thinking about the bridges between biological systems and electronics, we couple thoughts of smart devices and materials with the fields of biotechnology and nanotechnology. We think of biotechnology as the systems that tie biology to technology. For example, an implant that regulates blood chemistry is a biotech device. Nanotechnology refers to microscopic systems; *nano* itself means a billionth, so a nanometer, the size used to measure these microscopic systems, is one-billionth of a meter. Something created with nanotechnology need not be fused with biological systems; it need not exist in our bodies, for example. Someday, we'll have nanotech systems embedded everywhere: microscopic, interconnected, and widely distributed networked systems that exist in our walls, shoes, hats, pillows, and transportation vehicles, as well as in our very flesh.

In terms of biotechnology, neurotrophic electrodes are already in use. Says futurist Michael Zey,

> Other diseases and disorders are being treated with neural implants in what is being called "deep brain simulation" therapies. Doctors have achieved some success in using such implants to reduce tremors associated with cerebral palsy, multiple sclerosis, and other tremor-causing diseases. They are doing this by implanting electrodes in a section of the brain called the ventral lateral thalamus.[10]

A paralyzed man with an implanted neurotrophic electrode can now communicate with a computer system. The electrode, devised by Emory University Hospital scientists, is coated with chemicals that stimulate connections between the implant, surrounding nerve tissue, and the man's brain. Electrical signals run through the connections, and the electrode transmits them to a receiver on the surface of the man's scalp. He requires no wires running from his scalp beneath his skin. Remember that this man cannot physically move. However, using the neurotrophic electrode, he is able to transmit signals from his brain to a computer. He moves the cursor on the screen, selecting letters to spell words, and it is hoped that he'll soon be sending e-mail.[11]

The *New York Times Magazine* reported that in 1996 scientists received FDA approval to create brain-computer interfaces in almost-dead patients. Ideal patients were those without any motor control and suffering from brain stem damage—similar to the paralyzed man just described. The first operation was in 1998. By imagining he could move his paralyzed left hand, the first patient triggered an increase in his brain's electrical impulses, and these impulses were transmitted to a receiver on his pillow. The receiver translated the analog brain signals into digital computer signals and passed the digital results to a computer. Using this method, the patient could move a cursor on the computer screen.[12]

As for nanotechnology, early forms of these devices and their applications may be available soon. However, it seems unlikely that in the immiment future we will have nanotech machines coursing through our veins, cleansing us of fat; or nanotech machines in our brains serving as cyberpunk computers, possibly downloading our minds onto Crays or uploading the encyclopedia of the world into our skulls. The reason some experts predict that we're on the cusp of nanotechnology may be due to slight differences in how we define the term.

James D. Plummer, professor of electrical engineering at Stanford University, notes the following:

> Perhaps the broadest definition of a "nanostructure" is something which has a physical dimension smaller than 0.1 micron, or 100 nanometers (billionths of a meter). . . . Based on current rates of development, people have projected that around the year 2005 companies will be manufacturing in high volume integrated circuits that have dimensions around 0.1 micron. . . . So if size alone defines nanotechnology, then everyone of us is seeing practical benefits of it today.[13]

However, Plummer also points out that nanotechnology usually connotes nanosize machines. Eric Drexler, the author of the groundbreaking *Engines of Creation*, defines nanotechnology to include the following:

> Molecular manufacturing, that is, products can be assembled on the nanometer scale with extreme precision allowing the rearrangement

of individual atoms and molecules; materials with novel and/or adaptable properties, controllable by molecular manufacturing; miniaturization of electronic and mechanical parts down to the atomic scale, leading to nanomachines, whose compactness and efficiency could even outperform the cellular systems.[14]

Plummer agrees and further states that it will be at least twenty-five years before we have nanotech machines cleaning fat from our arteries.[15] *New York Times Magazine* predicts that we'll have to wait about thirty years.[16]

Nanotechnology research is booming. On April 27, 2001, *Wired* reported, "The quest for nanometer-scale computing is now a quantum leap closer to reality. In Friday's issue of the journal *Science*, physicists from IBM's Thomas J. Watson Research Center announce their fabrication of the world's first array of transistors made from carbon nanotubes."[17]

There's a reason why carbon nanotubes are so exciting to today's scientists. Early computers used switches that were actually vacuum tubes—large glass tubes in which electric current passed freely between metal wires. A binary on was when electrons were flowing in the tube. A binary off was when they were not. Later, semiconductor transistors took over the work of the vacuum tube. As a result, our computers became smaller and much more powerful. Today, experts believe that carbon nanotubes of rolled-up graphite may replace transistors. These nanotubes measure only a few nanometers in diameter.[18]

Other areas of nanotech-related research are also pushing forward at a rapid pace. In November 2000, *Wired* reported that major work was occurring in the field of "nanotechnology—also known as *bio-micro-electromechanical systems, or bioMEMS*—as a means to deliver drugs, supplements and therapies to specific sites in the body or to draw out a dosage over weeks, months or years."[19] BioMEMS combines biology and genetics with advances in computer and electronics technologies. For example, researchers are investigating techniques that combine robotic heart surgery and microchip technologies. Dr. Robert Michler, director of research and the chief of cardiothoracic surgery at the University Medical Center at Ohio State University, explains that his team coats microchips with chemicals, such as insulin, gene therapies, and heart medicines. Human clinical trials will begin in five years, and his team is already creating chips in hopes of using robots to insert them into animal hearts. As Professor Albert Pisano, director of the Electronics Research Laboratory at the University of California at Berkeley, states, "It's already science fact, not science fiction."[20]

For Doctor Octopus, the implications of the research are clear. Doctor Octopus's steel tentacles are somewhat like prosthetic devices. They are robotic machines that are hooked into a harness that he wears. Supposedly, this harness is grafted to his body. With this in mind, it makes sense that the harness can control the steel tentacles. The entire unit—harness plus tentacles—could be composed of smart materials, microdevices, and bioMEMS. Computer circuitry in both the harness and the tentacles would provide sensors, movement, and control. And current research further indicates that Doctor Octopus could indeed control the computer circuitry in the harness-tentacles apparatus using his mind. If we can already create brain-computer interfaces in paralyzed and almost-dead patients, enabling these people to control computer systems, then Doctor Octopus could control his limbs in a similar way.

Now, could the harness be grafted into his body? Again, given current research and projections into the near future, we see no reason why this would be impossible. If people's bodies can now support sophisticated prosthetic devices that include electronics and mechanical parts, then Doctor Octopus's body can support the same thing. We might suggest that his prosthetic device—the harness—be made a bit more aesthetic, however. After all, if he's going to marry Aunt May (as he does in a couple of comic issues), he should look his best.

Final questions about these robotic limbs: Could Doctor Octopus run at 50 miles per hour on his steel tentacles? Could he whirl them like a fan? If they are truly mechanical instruments, then we say: why not? Sure, it's plausible.

As for the radioactivity element—a standard component of superhero and supervillain comics—it does not mutate human beings; rather, it kills people. The radioactivity in Doc Ock's lab wouldn't graft the robotic arms to his body, creating brain-machine interfaces along the way and triggering criminal insanity in him. More likely, a high level of radioactive exposure would fry him.

But we're willing to overlook this flaw in the case of Doctor Octopus. Because the rest of him—the robotic limbs, the brain-computer interface, the prosthetic harness—is all plausible, as well as a lot of fun.

7

Leapin' Lizards
The Lizard

Meet **Erik Sprague,** the world's first human lizard. As reported on the *Ananova* Web site on May 12, 2001, Erik has been remodeling his body to resemble a human reptile. Sprague freely admits his inspiration comes from the supervillain, the Lizard, as depicted in *Spider-Man* comics.

As of May 2001, Sprague has had surgery to create a forked tongue and has horned ridges implanted in his skull. His fingernails are shaped like claws and he has scales tattooed over his body. He's undergone four hundred hours of tattooing and estimates he has another two hundred to go.

Erik, at twenty-eight, is working on his Ph.D. at the University of Albany in New York. He tours with the Jim Rose circus. He says he'd like to have a tail, but would only settle for one with real tissue. The only available tails are prosthetic ones.[1]

The source of Erik's inspiration is a comic book character named Dr. Curtis Connors. He first appears in *Amazing Spider-Man* (#6, November 1963). He's a good man, a decent man, a man who lost an arm helping his fellow men. There's not an evil bone in his body. Yet Dr. Curtis Connors is a tragic hero with one fatal flaw. It's the same flaw that doomed the most famous of all tragic heroes, Macbeth. Ambition. Fortunately, unlike Macbeth, who was pushed to his doom by the three witches, Connors has a guardian angel—the young man known as Spider-Man.

In "First Lizard," we're introduced to Dr. Connors, a surgeon who lost an arm while serving in the war (Korea, we assume, as Vietnam was still years in the future). After the war, Connors meets and marries his wife, Martha, and they have a son, Billy. Haunted by the loss of his right arm, he is dedicated to finding a serum that enables humans to regenerate limbs.

Connors moves his family to Florida's Everglades, so he has a constant supply of lizards to use in his experiments. After years of research, he finally comes up with a potion that he believes will allow people to regrow lost arms and legs. He tries it on a rabbit with complete success and no ill effects. Connors, in typical comic book fashion, doesn't think to share the news of his startling discovery with other scientists. Comic book scientists don't believe in teamwork or sharing the glory with others, which results in numerous discoveries being lost when the only copy of a special formula is destroyed or the researcher is killed. Instead, in the ultimate test of the serum, Connors drinks it himself to see if it will work on humans.

Doctors and scientists using themselves as human guinea pigs aren't just in comics. For hundreds of years researchers have exposed themselves to dangerous diseases or experiments. Marie Curie's hands were covered with radiation burns for most of her life. She died from radiation-induced leukemia in 1934. Many people who worked with X-rays died from this disease.

More recently, Stephen Hoffman, a member of a team of researchers searching for a vaccine to prevent malaria, injected himself (as did six of his coworkers) with an experimental drug, then let himself be bitten by malaria-infected mosquitoes. Hoffman came down with the disease, but it was more than a day and a half before the malaria was spotted in his blood and treated. Undaunted by his experience, he went on to invent a more effective vaccine. When asked why he had used himself as a human test culture, Hoffman replied, "Someone had to. Why not one of the people who helped develop the vaccine?"[2]

Unfortunately, not all doctors and scientists are so ethical. In *The Plutonium Files*, the horrific experiments of American atom bomb researchers conducted on an unsuspecting public during the 1940s and 1950s is revealed in chilling detail. None of the subjects were aware that they were being used in experiments dealing with high levels of radioactivity. Massachusetts Institute of Technology researchers fed radioactive oatmeal to children attending a state boys' school outside Boston. Prisoners in Washington and Oregon were exposed to blasts of direct radiation. Medical patients thought to have terminal illnesses were secretly injected with large does of plutonium to see how it affected their chances of recovery.

Fortunately for Spider-Man, Dr. Connors is an ethical man and only injects the serum into himself. The cure works, and Connors regenerates (rather quickly) his lost arm. But the changes don't stop there. Though nothing more happens to the experimental rabbit, the potion evidently affects humans differently. Connors is turned into a six-foot-tall lizard with thick green scales and a long alligator-like tail. Conveniently, despite the fact that his entire head is reshaped and mutated into that of a lizard's, he can still talk fine.

Connors runs into the swamp, leaving his wife and son to worry if he will ever evolve back into the wonderful guy he once was. Meanwhile, tourists in the Everglades find themselves confronted by a man-size talking reptile who claims the swamp for his own. Bullets don't harm this green monster, who rips trees to pieces when he gets angry. Within hours, Spider-Man is off to Florida to confront this new menace.

This being a comic book plot, it's not enough that Connors has devolved into an enormous lizard with only a foggy memory of his previous life. Adding an extra layer of menace, we soon learn that the transformation scrambled Connors's brain. The first of a new species of Lizard Men, he now plans to rule the Earth with the help of the world's many reptiles. Somehow, the miraculous growth potion has given him telepathic control over all the snakes and crocodiles in the swamps. Moreover, the serum has given the mutated scientist super-strength. Not that normal lizards are noted for being extremely strong, but it makes better drama for Spider-Man to fight a foe with enormous muscle power.

After a few minor battles involving the talking Lizard Man and his horde of obedient crocodiles, Spider-Man returns to the Connors home determined to find an antidote to cure the poor doctor. Working hard and using the notes of Dr. Connors, our high school hero and all-around scientific genius develops a potion that will change the demented experimenter back to normal. During a fight in an empty well, Spider-Man pours the fluid down the Lizard's throat, and within a few minutes the monster changes back into the good doctor, unfortunately losing his regrown arm in the process. Connors decides there are some secrets that humans aren't meant to know, and in true horror movie style he burns his research notes. The truth about the Lizard will remain a secret forever.

Of course, the best-laid plans of spiders and lizards often go astray, as Dr. Connors discovers when the cure is unstable and every few months he reverts back to lizard form. In one story, he haunts the New York City sewer system, an editorial nod to the urban myth of alligators in the sewers. In several adventures, the Lizard helps Spider-Man fight other animal-influenced villains (like the Rhino). Still, most of the time, Connors in lizard form usually schemes to turn everyone else in the world into a reptile, with himself as chief lizard. Over the course of forty years, he's remained a thorn in Spider-Man's side, appearing as a villain in well over sixty adventures.

The story of the Lizard gives rise to an obvious question about Spider-Man's reptilian foe. Is there any shred of truth lurking in Dr. Connors's experiments? Can we actually learn how to regenerate lost limbs based on what we learn from lower forms of life? Or is it all comic book pseudoscience?

As is often the case with comic book stories, we need to separate the illogical from the unscientific. We first must examine the plot and see what makes sense. Then, with the cliches stripped away, we can take a clear look at the scientific principles involved and hopefully reach some conclusions.

One of the most common plot devices used in the story is that Dr. Connors works by himself in a home lab in the middle of nowhere, or close enough, in the Everglades. Other scientists aren't around to ridicule his efforts, and unscrupulous businesspeople can't try to steal his discoveries—or so we assume, since businesspeople in comics are uniformly evil and greedy.

Unfortunately, Dr. Connors is also separated by those who could assist him with his research or save him when his work turns dangerous. One of the most interesting points made by the story is that it's Spider-Man acting not as a superhero but as a scientist (remember, he's just a high school biology student) who comes up with the antidote for the regeneration serum taken by Dr. Connors. In essence, this minor subplot reinforces the idea that while scientists spend months, sometimes years, developing important serums and medications, a high school student can duplicate their efforts in a few hours. Superhero comics often seem to regard science as something that can be done in a hurry.

If Dr. Connors had carried on his work in a well-equipped laboratory with numerous assistants and associates, the results of his experiments would have been quite different. He wouldn't have been forced to use the serum on himself as a guinea pig. A long, detailed study of the serum would have been conducted under strict supervision to make sure the drug made the necessary changes without harming the subject. Again, why the hurry to test the formula? After being without a right arm for ten years, couldn't Connors have waited a few more weeks before swallowing his potion? Besides, hadn't Dr. Connors ever watched horror movies featuring mad scientists? He should have known better than to experiment on himself with no one nearby to help.

Dr. Connors should have realized that burning his research notes at the end of his lizard adventure was a terrible mistake. Using the basic elements of his research, other scientists might have discovered a safe way to regrow limbs. Connors thought he was doing the right thing, but he was actually abandoning thousands of people without limbs. In a deeper sense, the true meaning of the story was not that scientists should avoid conducting dangerous experiments but that no one should be allowed to make decisions that will affect the health and well-being of thousands.

If Dr. Connors had acted logically, he would have stayed in the big city where he could have worked at a university with a government grant and gotten all the lizards he wanted for research without tramping through the Everglades. The entire adventure, as described in the comic book, could have taken place pretty much as described, other than possibly the Lizard claiming the school swimming pool as his domain instead of the swamps. Spider-Man still would have emerged as a hero, though he might have gotten some help formulating the antidote from other scientists on campus.

When "First Lizard" originally appeared, the entire concept of limb regeneration was the stuff of science fiction. Now, however, incredible strides have been made in biology. If Dr. Connors were working today, would he still be the only one investigating the subject? Or would he be just one of many scientists trying to discover how to regrow human limbs? Was this just another wild and impossible idea thought up by Stan Lee for a *Spider-Man* story? Or did Stan actually predict where science was going long before it happened? Amazingly, the truth is somewhere in between.

Forty years ago, Dr. Connors studied lizards in an effort to discover how they were able to regenerate lost limbs. After much research, he came up with a potion that he drank, very much in the style of mad scientists. Rewriting the story today, Connors would most likely use some sort of gene therapy to mutate the area he wanted to grow back. It's still science fiction, but the concept is getting closer to reality every day.

Scientists throughout the world are studying how amphibians and other animals regenerate limbs with the hope of someday using that knowledge to help humans do the same. One of the best-known researchers in the field is zoology professor Steven Scadding, who has worked on the mystery for more than a quarter of a century. Scadding primarily deals with axolotls, a species of salamanders that possess this regenerative power.

Scadding has discovered that a substance called retinoic acid (now available by prescription and all the rage among those trying to look younger), a form of vitamin A, is one of the main factors in regeneration. The acid seems to "switch on certain genes, triggering specific growth patterns."[3] Different amounts of acid in the blastema (the skin covering that grows over the stump of the removed limb) regulate how the limb grows back for axolotls. Scadding is investigating how the retinoic acid somehow allows various cells to communicate with each other and grow in the proper place.

Scadding hopes to learn how axolotls regenerate and apply that knowledge to human beings. It's a goal that's not very different from Dr. Connors's goal. "If amphibians can do it, and they have an arrangement of bones, muscles and cells similar to ours, there should be the potential for humans to do it, too," says Scadding.[4]

The zoologist is not the only one who is working on limb and nerve regeneration. Professor Lyn Bealsley, head of the Department of Zoology at the University of Western Australia, is leading a team trying to discover how simple animals regrow nerve fibers. The goal is to discover a cure for damaged nervous systems in human beings. If the research team succeeds, people like Christopher Reeve will someday be able to walk again.

Some nerves in our body, including those in our fingers and hands, do grow back when cut. Unfortunately, that's not the case with the nerves in our central nervous system. Those nerves try to regrow for a few days and then stop. No one is sure why. Dr. Bealsley and her team have been studying animals to see if they can learn the answer.

Lizards were among the first animals selected for limb regeneration, demonstrating that Dr. Connors was working in the right direction. A lizard can have its optic nerve smashed and the nerve fibers will regrow from the brain to the eye within six months. However, that growth isn't always perfect. Sometimes, the nerves grow to the wrong section of the brain. The lizard is capable of regrowing nerve cells but not always to the right location. Which prompted Dr. Beals-ley and her team to look at an even simpler animal than a lizard: a frog.

Frogs regrow nerve fibers to the right place in the brain, but it takes them much longer to perform this feat than it takes lizards. Not satisfied with the time factor, Dr. Bealsley went searching for another animal that could regrow nerve fibers quickly. After much looking, she discovered goldfish.

Crush the optic nerve of a goldfish, and immediately hundreds of thousands of optic nerve cells begin rebuilding the path between the severed eye and the brain. In approximately one month, the goldfish's 20–20 vision is completely restored. The small fish possesses the right molecules to duplicate nerve cells and the chemical guideposts to direct the nerve growth to the proper area of the brain.

While Dr. Bealsley and her team haven't yet discovered exactly how these special molecules and chemical guides work, they have plenty of test subjects to study and have high hopes of someday being able to apply what they learn to the human nervous system. Maybe Dr. Connors's dreams weren't that unbelievable. Perhaps he just picked the wrong animal to study, although Spider-Man fighting the Incredibly Ruthless Goldfish might have been a hard sell.

Which leads us with just the slightest stretch of the imagination to the planarian. For those readers who have forgotten their high school biology, that's the name of a quarter-inch-long worm that can live in the ocean, in lakes, and in the earth without any problem. It's also a popular subject for experiments in sophomore biology classes, because when you cut off its head, the planarian has the enviable tal-ent of being able to grow it back.

Some planarians, which are also known as flatworms, reproduce in a unique manner. A planarian stretches until its body pulls apart into two pieces. The section with the head grows a new body, while the sec-tion that's all body grows a new head. Interesting stuff, but flatworms and humans are far apart on the evolutionary scale. We're very distant cousins. Yet distant in biology is closer than most people realize.

According to Dr. Sanchez Alvarado, a researcher at the University of Utah Health Sciences Center, 70 percent of the 4,500 genes that researchers have studied in planarians are found in people.[5] By studying planarians, Alvarado hopes to learn the processes behind regeneration and apply them to human medicine.

Planarians are metazoans—that is, they're animals that originate from a single cell and grow into complex organisms with cells arranged into different organs. They're one of a number of metazoans that regenerate. Salamanders are another.

A tissue sample as small as $\frac{1}{279}$ of an inch can be taken from just about any part of a planarian and will regenerate the entire worm. That's comparable to a person losing a finger in an accident and the finger growing an exact duplicate of the victim.

Scientists suspect the secret of a planarian's regeneration lies with the vast number of cells called neoblasts contained in its body. Neoblasts can transform into whatever cells a planarian needs. They can become neurons if the planarian needs neurons, or they can become muscle cells if it needs muscle cells. Scientists aren't sure how many neoblasts are required to regrow an entire worm. Dr. Alvarado thinks one neoblast might be enough to do the job. "In principle," says Alvarado, "they [flatworms] are immortal."[6]

If all this sounds vaguely familiar, it's because neoblasts have been in the news during the past few years. In humans, they are usually called stem cells.

In the 2004 version of the "First Lizard" adventure, Dr. Curt Connors might be pursuing the same goal as real-life researchers Dr. Lyn Bealsley and Dr. Sanchez Alvarado. They're all looking for cells that will regrow body parts that have been destroyed or severed. Instead of trying to replace his arm, Dr. Connors might be working on a serum to make a quadriplegic walk or help Alzheimer's patients regain their memories. Forty years ago, Dr. Connors's experiments seemed impossible. Not so today, where they are on the cutting edge of science.

8

Clothes Make the Man
Venom

It all started with a huge Marvel Comics crossover event called *Secret Wars*. A large cast of superheroes and supervillains were transported to a planet called Battleworld by a near omnipotent being called the Beyonder and forced to fight one another. In part eight of the multi-issue story, Spider-Man found himself in dire need of a costume. His regular outfit had been ripped to shreds, and he had no way of repairing it or making another. However, he had noticed that Thor, the God of Thunder, had reached into a machine and replaced his torn cloak and smashed helmet with exact replicas. Deciding to take advantage of the system, Spider-Man did the same thing. But he reached into the wrong machine. Instead of pulling out a duplicate costume, Spider-Man withdrew a small black ball. Instantly, the material fanned out over his entire body, covering him from head to foot in a costume that looked quite similar to the outfit of Spider-Woman.

This new black-and-white costume was a complete change from Spider-Man's original outfit. According to editor Jim Salicrup, "The big deal was that never before had a superhero as well known and successful as Spidey changed his appearance. Up until that point, only second-string characters would don new duds. It was a way to try boosting both reader interest and sales by giving characters a new look." Salicrup said the costume change was so astonishing that it was covered in newspapers and TV news programs. However, the change wasn't done entirely to attract attention. It actually forwarded an entirely new and unusual plot.

When he finally returned to Earth from Battleworld, Spider-Man discovered that his new form-fitting outfit increased his strength, could transform into any outfit he imagined, and had its own web-shooters and an unlimited supply of webbing. It was the perfect superhero costume except for one thing: it was alive.

The costume was more than a costume. It was an alien symbiote that took on the appearance of a costume and fed off of Spider-Man's emotions. When Spider-Man finally realized there was something strange about his uniform, he brought it to Reed Richards, the leader of the Fantastic Four. Richards discovered that the costume was alive and trying to meld itself to Spider-Man's body and mind.

Reed Richards succeeded in separating the costume from Spider-Man by shooting it with a sonic blaster. He imprisoned the suit in a special chamber, but it escaped and returned to Spider-Man's closet, where it disguised itself as Peter Parker's old costume. When Parker donned the costume, he immediately realized he was once again wearing the alien symbiote. Knowing the being's vulnerability to sound, Spider-Man went to the bell tower of Our Lady of Saints Church. The noise knocked out Spider-Man but also seemed to destroy the alien costume. When Spider-Man awoke, the alien uniform was gone. Spider-Man, however, was wrong to think it was dead.

Though badly hurt, the symbiote slithered through the walls of the church until it came upon Eddie Brock praying at the altar. Brock had been a successful newspaper columnist at the *Daily Globe*. In what seemed to be an incredibly lucky break, a man named Emil Gregg contacted Eddie and confessed to being the serial killer the police had dubbed the Sin-Eater. Using information supplied to him by Gregg, Eddie wrote a series of columns about the crimes that made front-page news and spiked the *Globe*'s circulation. That was until Spider-Man caught the real Sin-Eater, Detective Stan Carter. Brock was branded a fraud by his fellow newspaper reporters and fired from the *Daily Globe*. Needless to say, he blamed his troubles on Spider-Man.

Trying to reduce his anger, Brock started working out, turning his body into that of a star athlete. But nothing could lessen his anger with Spider-Man. Finally, consumed by anger and despair, Brock went to pray at Our Lady of Saints, planning afterward to kill himself. It was in this heightened state of emotional distress that the symbiote discovered him. Brock and the alien costume merged, both obsessed with destroying the crime fighter who had caused each of them so much pain: Spider-Man.

The alien costume worked well for Brock, giving him all the powers of Spider-Man, including superstrength and the ability to block Spider-Man's spider sense. It also revealed Spider-Man's secret identity to him. Eddie added a maniacal grin to the black uniform, and together he and the symbiote took on the name Venom.

Venom soon became one of the most popular Spider-Man villains. As such, within a fairly short time he was given his own miniseries, where he was able to manifest a heroic as well as an evil personality. After all, both Eddie and the symbiotic costume hated only Spider-Man, so it was okay for them to help innocent bystanders. This dual nature of being both evil and good helped make the character even more popular. Even when confronting Peter Parker's wife, Mary-Jane, Venom threatened but did not harm. He was only out to destroy Spider-Man and refused to hurt anyone other than his nemesis.

Marvel Comics of the 1990s could almost be labeled the Venom years, as the character appeared in so many miniseries, guest star roles, and spin-off adventures. Venom was so popular that he (literally) spawned a second symbiote called Carnage. Bonding with a serial killer, Carnage possessed all of Venom's powers but was pure evil. He served as a high-powered villain for both Spider-Man and Venom.

Without a doubt, Marvel writers Tom DeFalco and Roger Stern hooked a generation of comic readers with their concept of a suit of clothes that could think for itself. It was a unique idea that had been tried once or twice in comics but never with such imagination and success, leaving us to ask, how possible is such a concept in real life? Is Venom only possible as an alien symbiote, or are scientists (and fashion designers) even now working on clothing that can think? The answer is a place where comics and the near future collide with astonishing results.

Imagine a jogging suit that tells you how to exercise. The suit keeps track of your heartbeat. It plays music from an integrated MP3 player and adapts the rhythm and beat to make you run faster or slow down. The built-in mobile transmitter sends information about your progress to your health club, and the readout is available by the time you finish your shower. It's also handy since the electronic ribbon that serves as a transmitter can be used for wireless communication. Just in case you get lost or jog into an avalanche.

Because you want comfort as well as convenience, your jogging suit is lightweight and doesn't need any batteries—ever—to run these devices. It generates power from the differences in your body heat and the surrounding air. And since you like to look good while you're running, you've programmed the interactive textiles to change colors every few minutes.

Impossible? An alien symbiote? Not according to the giant Dutch corporation Philips Semiconductor. Scientists working for their lab in England have already made a ski jacket with a built-in thermometer that monitors the user's body temperature. If there's a sudden change, the jacket's fabric heats up. A global positioning satellite system is built into the jacket to locate a skier, if necessary. Plus, a proximity sensor on the back of the coat notifies the skier if any other skiers come too close.[1]

Philips also manufactures a denim jacket with a collar that contains stereo speakers. With the collar turned down, people a few feet away can hear the music. With the collar up, only the owner hears it. The jacket system receives phone calls and can download music from the Internet. Even Venom can't do that.

Novelty items that will sell at Christmas and soon be forgotten? Not according to Dr. Maggie Orth, cofounder of International Fashion

Machines in Cambridge, Massachusetts. "It's a much different way of thinking about a digital or computer medium," says Orth, a graduate of the Massachusetts Institute of Technology's Media Lab. "Electronic textiles still are at a 'black art' stage. But this industry is in a growth period."[2] According to Orth, some of the computer/clothing will be hitting the shelves during the next few years.

It's a field—variously known as smart fabrics, e-textiles, wearable computers, or intelligent textiles—that many futurists anticipate will become one of the hot drivers of the American economy. Scientists expect it to propel technology forward in general, because its applications are so diverse.

"Society in the next 10 to 15 years will involve people being surrounded by electronic gadgets with ambient intelligence," says Werner Weber, senior director of corporate research and emerging technologies at Infineon Technologies AG of Munich, Germany. The firm is developing electronics to be used in smart textile applications for consumers. "The wearable electronics will be woven in, so customers don't have to think about manuals."[3]

Obviously, jogging suits aren't for everyone. Instead, consider shirts that monitor your vital signs and signal to your doctor if your heart starts beating too fast or your blood sugar is too low. The U.S. Navy funded a project in 1996 that eventually turned into the Smart Shirt, a product commercialized by SensaTex Inc. in Atlanta, with technology from Georgia Tech Research Corp. The T-shirt functions like a computer, with optical and conductive fibers integrated into the garment. It can monitor the vital signs, such as heart rate and breathing, of wearers, including law enforcement officers, military personnel, astronauts, infants, and elderly people living alone. In a few years, such shirts are going to be the accepted wear at retirement villages across the United States.

Of course, not all elderly men and women wear T-shirts. That's why Philips Research Laboratories has developed bras, vests, pants, and underwear that measure and analyze the subject's heartbeat. These special sensors are linked by wireless technology to a customized phone that automatically contacts a hospital (or ambulance service) in case of emergency. According to research director Karl Joosse, "They are so easy to use, the patients wouldn't realize they are wearing special underwear."[4] Philips expects to market their new clothing line within the next two years.

The technological possibilities of smart clothing are of great interest to the U.S. military. The armed forces have been experimenting for years with weaving computer and communications technology into uniforms. Future combat wear will keep soldiers warm, fight germs, and eventually detect and fight chemical agents. One scenario has soldiers wearing vests that can instantly detect when a trooper has been shot, where on his body the bullet hit, and how far it penetrated. The generated information would be sent in seconds to the nearest medical team along with the wounded soldier's position.

Much of this clothing research for the soldier of the future is being conducted at the U.S. Army Soldier Systems Center in Natick, Massachusetts. Scientists and technologists are tackling a variety of textiles that can transport power and information. One example is a soldier sticking his or her intelligent glove finger into water to see if it is safe to drink. The soldier could communicate with others by a fabric keyboard sewn or woven in as part of the uniform's sleeve.

Camouflage is an important part of modern warfare. However, as warfare moves from the jungles to the cities, blending into the environment has become a lot more difficult. People aren't chameleons. The only perfect camouflage outfit is one that makes its wearer invisible. And that's impossible. Or is it?

Not according to Professor Susumu Tachi of Tokyo University in Japan, who is the leading proponent of optical camouflage.[5] Tachi's system cloaks a soldier in a reflective silver coat. Miniature video cameras on the back of the coat photograph what is behind the soldier and transmit those images to a projector several feet in front of the soldier. The projectors beam the images onto the soldier's coat. Anyone looking at the soldier sees what's behind him. Thus, the soldier blends perfectly into the scenery. The biggest problem with

Professor Tachi's system is that the soldier always needs to be standing in front of a video projector. Still, it is a beginning in what will most likely be the next major advance in camouflage technology.

Possible future army gear includes global positioning systems, identification sensors, monitors, chemical detectors, and computer-controlled weapons. However, putting such devices into smart uniforms isn't easy. "The goal is to provide the soldier with executable functions that require the fewest possible actions on his or her part to initiate a response to a situation in combat by using intelligent textiles," says James Fairneny, a project manager at the Natick lab.[6]

Fairneny's group is working on ways to make electronic equipment that's integral to textiles. Latest projections have such smart uniforms five or six years away from field use.

Another possible area of huge growth in smart clothing involves antibacterial and antimicrobial polymers. Gregory Tew, assistant professor in the department of polymer science and engineering at the University of Massachusetts, Amherst, and his colleagues are devising molecules that act in much the same way as cells in the human body to combat germs. In addition to embedding such molecules into clothing, they could be put into paints and plastic coatings. Antibacterial polymers could be used in everything from socks to children's clothing to hospital surgical gowns. Such material could even be used on kitchen countertops and refrigerators to fight germs.[7]

Science fiction? Comic book stuff? It's all coming, says Dr. Sundaresan Jayaraman, professor of textile engineering at Georgia Institute of Technology. So are shirts that will signal your lights, air conditioner, and CD player to turn on upon your arrival to your home. "Our goal is to make clothing as easy to use as a microwave oven," says Jayaraman.[8]

You might think you're not ready for clothing that slides onto your body and spins organic spider webbing, but that day might be a lot closer than you think. And it won't only be in a few isolated areas in the United States.

Take a look at Europe. A very strange event took place in England a few years ago. For the first time since 1951, ten thousand volunteers had approximately 130 points of their body measured in department stores and fashion shops. This project was used to determine whether people in England had become taller, fatter, and wider over the past five decades. It was part of an ongoing study in body evolution, but it wasn't conducted to prove that the world was getting fatter. Instead, it was to help design clothes for the twenty-first century. This idea was inspired by a new hybrid science called evonetics.

Invented by Professor Wolf D. Hartmann of the Klaus Steil-mann Institute (KSI) for Innovation & Environment, in German evonetics stands for *evo*lution of *net*works and energies for design of technology and interfaces for customer use. In simpler terms, evonetics means the merging of technology and design to develop multifunctional clothing for the future.

Evonetics foresees the use of nanotechnology and biotechnology, microsystem technology, and microscopic computers as a major link between science and clothing. Hartmann and many other European scientists see technology blending with new and emerging lifestyles to create a future world unlike anything we've experienced so far. It is through their efforts that KSI and a number of other high-tech companies and institutions have formed a European network of fashion industries and high-tech manufacturers known as ICEWES. Scientists working for this powerful new conglomerate are already researching the symbiosis of human needs and technical systems to form a "fusion of nature and technology, art and environment."[9]

Meanwhile, in the United States, MIT and Georgia Tech are also working hard on developing wearable computers. The fashion industry is already gearing up to new developments like intelligent textiles, I-wear, and fashion engineering. Wearable electronics, embedded technologies, and disappearing computers are the hot buzzwords for development of this new clothing and its marketplace. Applications for the health and security sector, such as the smart shirts mentioned earlier, are seen by manufacturers as one of the major clothing trends over the next ten years.

Speaking at the Technical University of Brandenburg, Germany, in the first series of lectures on evonetics, Professor Hartmann stated that "Darwin's Survival of the Fittest will become, in the network age, Survival of the Webbest."[10]

Not everyone, however, is as convinced as Professor Hartmann of the importance of networking. In March 2003, Benetton, one of the world's largest clothing manufacturers, announced that it was planning to put radio frequency ID (RFID) chips manufactured by Philips Semiconductor in all of its new clothing. The chips are designed to cut the cost of inventory control and can only be tracked within a few meters. Still, consumer groups worried that government agencies might use RFID signals to track citizens. A U.S.-based consumer group, Consumers Against Supermarket Privacy Invasion and Numbering (CASPIAN), called for a worldwide boycott of Benetton. Three weeks later, the clothing manufacturer declared that it wouldn't be using the RFID tags after all. Yet the technology exists, and sooner or later, locating people by what they are wearing will be as common as calling them on their cell phones.

The alien symbiote bonded with Peter Parker and increased Spider-Man's powers from superhuman to extra-superhuman. Later, it melded with Eddie Brock to make a normal man into Venom. However, Eddie only gained certain powers when he wore the symbiotic costume. That's not the case with smart clothes. Each outfit can be tailored for a different use. Venom might be a near unstoppable force in today's comic books. But wait twenty-five years and he might have a hard time defeating an ordinary person who owns a closet full of clothes.

9

Grodd Almighty
The Evil Super-Gorilla

A recurring supervillain in *The Flash* is Grodd, the Super-Gorilla. He belongs to a race of Super-Gorillas that have extraordinary genius, including the ability to hypnotize humans and other species via mental telepathy. In addition, their entire civilization—aptly named Gorilla City—is hidden from the world. It lies deep within the African jungle, where an entire city of Super-Gorillas thrives under the shield of a protective machine that cuts the city from human senses.

Grodd first appeared in *The Flash* #106 (April–May 1959). Subsequently, he was featured in a few dozen Flash tales with titles such as "Gorilla Warfare Part I," "In Grodd We Trust!" "3-Way Fight for the Super Simian," and "The Apes of Wrath." His origin story explains that something, possibly a meteor, crashed into the heart of Africa and showered a band of gorillas with rays. These rays somehow gave the gorillas superintelligence.

Later, the leader of Gorilla City, Solovar, was on an exploratory mission when humans captured him. Not wanting to reveal the existence of Gorilla City, Solovar acted like an ordinary gorilla and was placed in the Central City circus. The evil Grodd showed up in Central City and stole Solovar's secret method of controlling other beings using hypnotic mental telepathy. And that's how the Flash got involved: they spent years battling each other, as Grodd tried to conquer the other Super-Gorillas as well as the entire world.

In "Return of the Super-Gorilla" (*The Flash* #107, June 1959), we learn that Grodd has escaped from prison once again. Solovar calls the Flash using a vibration aura number that's based on vibration frequencies recorded by special devices. When the Flash arrives at Gorilla City to help Solovar find and capture Grodd, Solovar focuses the protective machine on the Flash to bring him into the gorillas' field of vibration. The jungle clears to reveal, as if by magic, a city the size of New York. (Another recurring plot device in the Grodd stories is that Solovar can also erase the memory of Gorilla City from the minds of humans.)

Meanwhile, Grodd uses hypnotic mental telepathy to force bird people to be his slaves. These bird people live on floating islands beneath the ground, and under Grodd's spell, they intend to rise up and take over the Earth.

Being an evil genius, Grodd has invented a devolutionizer ray that he plans to use on his gorilla civilization, turning the rest of them into ordinary primitive primates. To make a long story short, the Flash destroys the devolutionizer machine, punches Grodd's face about a hundred times, and returns the evil gorilla to prison. Repeatedly in *The Flash* comics, Grodd escapes, tries to take over the planet, and the Flash returns him to prison.

The main questions in Grodd comics are: How do we measure animal intelligence? Can we increase animal intelligence? Indeed, can a gorilla be a genius? Is there a method—for example, using a devolutionizer ray—to reverse evolution and reduce the population of Gorilla City to dumbfounded, primitive creatures?

Gorillas are great apes, which also include orangutans, chimpanzees, and bonobos. Humans, great apes, and the lesser apes (such as gibbons and siamangs) are all part of the superfamily called Hominoidea, which in turn is part of the suborder Simiae in the order Primates.

Scientists speculate that the earliest primates existed approximately 70 million years ago. The great apes split from the lesser apes approximately 20 million years ago. Genetically, the gorilla's closest relative is the chimpanzee. Human's closest relative is also the chimpanzee. Because of this parallel, scientists, comic book writers, and many other people have long been fascinated by the possibility of great ape intelligence.

Gorillas live in equatorial Africa: in Nigeria, Cameroon, the Central African Republic, Gabon, the Congo, equatorial Guinea, Zaire, Rwanda, and Uganda. Mountain gorillas are almost extinct. Approximately four thousand western lowland gorillas remain. Eastern lowland gorillas are faring better, with estimates of ten to forty thousand. Regardless of where they live, gorillas are shy, gentle, quiet vegetarians, and humans are their only enemies. Should gorillas obtain genius-level intelligence and become capable of designing protective shields against humans, it makes sense that they would do so.

So let's get back to the notion of gorilla intelligence. Most people think of intelligence in human terms. If we can prove that another animal—for example, a gorilla—has attributes of *human* intelligence, then we will consider that animal to be intelligent. Of course, this anthropomorphic way of defining potential intelligence in other animals really has no basis. Who are we to say that there is no other form of intelligence than our own? We humbly argue that any definition of a gorilla's intelligence should take into account the animal's environment, the stimuli to which it must respond, and the social organizations to which it belongs.

That said, human intelligence is often measured by the ability to compile facts and use logic to arrive at new and creative approaches to solving problems. Intelligent people can juggle multiple ideas, they're quick and witty, and their interests are versatile.

Humans plan ahead. Animals prepare for winter and know how to hunt for food, but they do not exhibit the elaborate plotting of humankind. Humans are deceptive and manipulative. Animals deceive to some extent but not nearly with the elaborate forethought and motivations of people.

According to theoretical neurophysiologist William H. Calvin, "Language is the most defining feature of human intelligence: without syntax—the orderly arrangement of verbal ideas—we would be little more clever than a chimpanzee."[1] Calvin explains that the early humans embellished the grunts and gestures of ape communication by inventing syntax. Chimps, for example, cannot add new words to their vocabulary. They are confined to approximately three dozen grunts along with gestures. They cannot combine three grunts to make one new word. Humans are also confined to approximately three dozen grunts, which we call phonemes. The difference is that combinations of our phonemes convey meaning. We are able to string together phonemes to create new words, new meanings, and new ideas.

While concrete findings are few, much research has occurred regarding animal intelligence. Specifically, recent research has focused on the intelligence of gorillas. For example, Dr. Francine Patterson, who heads the Gorilla Foundation, has been teaching sign language to a gorilla named Koko since 1972. Patterson's work is very impressive. And Koko is equally impressive. The gorilla knows more than one thousand words in American sign language, and she understands approximately two thousand words of spoken English. She initiates conversations with humans, and her sentences average from three to six words long. In addition, says Dr. Patterson, "All primates express emotions, but because of her command of sign language, Koko can convey to us feelings that her wild counterparts cannot."[2]

And Koko is not alone. Her gorilla friend, Michael, also understands spoken English and uses sign language to communicate. Michael has mastered more than six hundred words in American sign language.

Aside from speaking English and using sign language, large apes are also capable of reading and writing English. Sally Boysen, director of the Comparative Cognition Project at Ohio State University, has taught chimpanzees how to count, add and subtract, and think abstractly. Her research has shown that chimps have humanlike intelligence that isn't necessarily manifested in the wild. If the chimps live in more humanlike surroundings, then they demonstrate the ability to think abstractly, to do math, and yes, even to read and write.[3]

So if language is a strong measurement of intelligence, then it's quite possible that gorillas are much more brilliant than we assume. Koko's human IQ measures between 70 and 95. The average human IQ is 100.

Some scientists think that brain size is a good measure of animal intelligence. It is true that primates have larger brains than less intelligent creatures such as raccoons. And it is also true that primates have larger neocortices than other animals. Many scientists postulate that brain size and large neocortices are measures of an animal's intelligence.

Calvin explains, "Thoughts are combinations of sensations and memories. . . . They take the form of cerebral codes, which are spatiotemporal activity patterns in the brain that each represent an object, an action, or an abstraction." He estimates that "a single code minimally involves a few hundred cortical neurons within a millimeter of one another either firing or keeping quiet."[4]

The human cerebral cortex, which is used when solving novel problems, is 2 millimeters thick. If we were to flatten the extremely wrinkled cerebral cortex, it would be about the size of four sheets of 8½ x 11 inch paper. A chimpanzee's cortex would be about the size of one sheet. A rat's cortex would fit on a postage stamp.[5]

Other researchers conclude that brain size is irrelevant to intelligence.[6] For one thing, scientists tend to measure brain size by weight rather than taking into account the brain density differences among various animals. Brain density is calculated based on the proportions of white and gray matter, which has to do with the number of neurons per unit volume in the brain. This directly influences the information processing ability of the brain.

If gorillas are capable of achieving at least near-average human intelligence with training, can a gorilla become a genius by human standards? Given the status of current research, it seems unlikely. However, if we assume that highly intelligent gorillas have a need to communicate extensively with humans and that they can do so, then perhaps over extensive periods of time gorillas will evolve to have greater humanlike intelligence. Thus, the future of gorilla intelligence might depend on whether humans start treating gorillas as something other than dumb animals. It's a scenario right out of Poul Anderson's classic 1950s science fiction novel, *Brain Wave*, in which the Earth emerges from a cosmic cloud and the intelligence of all life on the planet suddenly ceases. Humanity ends up abandoning Earth to explore the galaxy, leaving gorillas in charge of the world.

According to the origin story of Gorilla City as related in *Flash* comics, rays from a meteor or a crashed spaceship increased the intelligence of a band of gorillas in Africa. We know from our chapter on the Incredible Hulk in *The Science of Superheroes* that rays, radioactive or cosmic, don't enhance living cells but destroy them. The Gorilla City origin story is about as believable as all the giant-insects-created-by-atomic-bomb-tests movies of the 1950s. Is there a more logical explanation for Grodd's intelligence? The answer to that question is evolution.

As in our previous book, we feel it only proper to mention that the United States is the only major industrialized country in the world where human evolution is not accepted as fact by a majority of the population. More than 45 percent of the people of our country do not believe in any type of evolution, though many of them seem to have little problem believing that dinosaurs once walked the earth. Fundamentalist religious leaders are often quoted as saying evolution is a theory, not a proven fact. The answer to such a remark is that gravity is only a theory. So is electromagnetism. Yet no one states that gravity is only a theory and not a fact. Gravity is accepted as true because we are not floating in the atmosphere. Electromagnetism is the reason our computers and TVs work. The evidence of this theory is all around us.

What about the theory of evolution? Can evidence of evolution be found in everyday life? Of course. Evolution explains why biology works, how cloning takes place, how genetic engineering is done, and a thousand other facets of daily life. Evolution is more than theory. It is fact.[7]

The most common mistake made about evolution is that it has come to an end. When fundamentalists question, "Why isn't evolution still taking place? And if it is, then why aren't gorillas still evolving into humans?" they're actually raising legitimate points. But they ignore that both questions have answers. Evolution is still taking place. Gorillas are still evolving. They're just no longer evolving into humans. If we want to believe *The Flash*, they're most likely evolving into Super-Gorillas.

One of the basic problems in comprehending evolution is the quite natural, egocentric (ego-species?) notion that humankind is the end-all result of the evolutionary chain of animal life. In charting the course of evolution on our planet, we always start with one-celled beings evolving into many-celled creatures evolving into invertebrates, then into fishes, amphibians, reptiles, dinosaurs, mammals, and finally resulting in humans. The progression through these various stages is correct. But the fundamental implication that the other species that form part of this chain either died out or stopped evolving is not. Evolution isn't a straight line but a multibranched tree in which the rise of humanity is just one of many branches. We might be the crowning achievement of evolution (not according to the dolphins), but we still are only one successful branch among many.

How does evolution tie in with Grodd the Super-Gorilla? Let's turn to the writings of Stephen Jay Gould, the brilliant scientist who died in 2002, for the answer: punctuated equilibrium.

For years, paleontology was considered one of the most static, boring sciences. Researchers spent most of their time uncovering fossils, labeling them, and reporting their findings in scientific journals. There seemed little else to study until Stephen Jay Gould and Niles Eldredge wrote an article in 1972 about punctuated equilibrium, which shook the paleontology community to its roots. A firestorm of criticism and debate upset the field as older, well-established scholars attacked the controversial new theory and its supporters, who modestly called themselves the Young Turks after the twentieth-century revolutionary party in Turkey.

In its simplest terms, punctuated equilibrium states that rapid speciation (the processes that lead to the creation of a new species) often takes place in small population groups that are separated, often by geography, from other members of their species. The evolution of this small group happens relatively quickly (over a few thousand years, a blink of the eye in the lifetime of our planet), and at the same time, due to the isolation and small size of the group, no fossilized remains are left. When the isolated group later rejoins the original population, the more evolved creatures from the small group quickly replace their less advanced cousins. Thus, the more widespread fossil records of the large group show this merging of two branches as an abrupt change in the species.[8]

Originally, this radical new theory seemed to contradict Darwin's belief in natural selection and the gradual changes that take place during evolution. However, close examination of Darwin's work has revealed that he never stated evolution had to take place over a long period of time and natural selection was only one of several processes that caused species to change. Though punctuated equilibrium has remained controversial, it has transformed paleontology from an unnoticed science into one of the most vibrant fields of scientific research.

In particular, punctuated equilibrium offers a valid explanation for the existence of Super-Gorillas. Following the main thesis of the theory, a group of reasonably intelligent gorillas would become isolated from all other primates in the jungle. An earthquake trapping them inside an extinct volcano might be the best comic book solution. The tribe would spend the next several thousand years surviving in this closed, somewhat safe environment. A chance mutation resulting in a gorilla capable of learning would occur, and with the narrow confines of their isolated world, this mutation would soon spread throughout the entire population. After a thousand years of interbreeding and continued learning, these new gorillas would finally figure out how to escape from their prison and return to the jungle. In the only variation from Gould's theory, the smarter gorillas would refuse to breed with their stupid cousins and would continue to remain an isolated tribe, this time by choice instead of circumstances. A few thousand more years would result in the development of Gorilla City. Unable to accept mere evolutionary processes as the reason for their intelligence, the gorilla civilization would invent a crashed spaceship as evidence of their destiny to rise above their fellows.

Speaking of evolution, we next ask if it's possible to reverse it using something like Grodd's devolutionizer ray. Could Grodd train his ray on Gorilla City and reduce the population to dumbfounded, primitive creatures?

The commonly accepted definition of evolution is a change in the gene pool of a population over time. A population is a set of individuals, such as a set of gorillas. Each gorilla has different traits. An individual does not evolve; rather, it keeps the same genes for life. So immediately we start wondering whether it's at all conceivable that Grodd could devolutionize his fellow gorillas. Since evolution deals with changes in a population over time, so-called devolution would imply the reversal of changes in a population over time. Devolution, as with evolution, would not occur immediately in individual gorillas.

The only explanation we see of Grodd's devolutionizer ray is that perhaps it kills brain cells. Perhaps it operates as some enormous form of electric shock–type treatment, turning gorilla brains into mush. Far-fetched, at best.

Without his fellow Super-Gorillas, Grodd's own seed would become extinct. There would be no lineage. Grodd could not mate and spawn future evil genius gorillas. So Grodd's choice to eliminate all other intelligent gorillas makes no sense purely from the perspective of an animal's instinct to prolong its line. Evolution occurs by combining the genes of creatures, and without superbrilliant gorilla genes to combine with Grodd's genes, gorillas would cease to evolve as creatures with humanlike intelligence.

As far as we know, colonies of Super-Gorillas do not exist. Of course, before Hiroshima and Nagasaki, atomic bombs didn't exist, either. Evolution does work, and it has worked for billions of years. While Grodd the Super-Gorilla hasn't made himself known to humankind, the world of science is full of surprises. Who knows what surprises, no matter how implausible they seem, we might discover?

10

A Magnetic Personality
Magneto

Powerful heroes need powerful villains. If a comic features a group of heroes fighting for truth and justice, usually there's an equally well-matched number of villains fighting back. There are, however, some mismatches in the Marvel Comics universe. The sinister Dr. Doom often fights the entire Fantastic Four in battles that the quartet seemingly cannot win. Nearly every hero in the Marvel universe fought Thanos, the rogue titan, in the world-shaking *Infinity Gauntlet*. The Dark Phoenix possessed the power to destroy galaxies when she struggled against her fellow X-Men. Still, it's rare that a villain is so powerful that he can't be defeated by a single superhero but needs the combined efforts of a team to do the job.

In the realm of the X-Men, there are a number of incredibly devastating villains. Powerful X-Men foes have included the already mentioned Dark Phoenix, Erik the Red, Holocaust, Apocalypse, Mr. Sinister, the Goblin Queen, Typhoid Mary, the Shadow King, and many others. But the most dangerous villain of all is the man called Magneto.

Magneto makes his first appearance in the initial issue of *X-Men* (#1, September 1963). Dressed in a red and purple uniform and a bizarre helmet complete with metal horns, he refers to himself as Magneto, Master of Magnetism. As with many supervillains, he enjoys talking to himself—not one of the best indications of sanity.

He also has the bad habit of referring to himself in the third person—a sure sign of an overwhelming ego.

Magneto strikes the army base at Cape Citadel, where NASA is testing its first rocket ships. Using his magnetic power, Magneto causes six launches to crash. It gives him great pleasure, demonstrating he can stop the space program in its tracks. Yet he doesn't claim credit for the sabotage, negating any publicity he might get for his actions. In these early *X-Men* adventures, pace dictates the story details instead of logic. After the last crash, in an attempt to startle and confuse the troops at the base, Magneto uses magnetized dust particles to spell out a huge skywriting warning: SURRENDER THE BASE OR I'LL TAKE IT BY FORCE!—MAGNETO.[1]

The base personnel are suitably shocked, but they don't surrender the base. Skywriting might be an impressive trick, but it's not very frightening. Magneto, however, seems surprised that the troops don't run screaming in fear from their posts. To make his point, he sets off one of the base's guided missiles to destroy some nearby deserted ships (killing innocent civilians in the early years of Marvel Comics was taboo). The soldiers still refuse to budge, which really annoys Magneto. Dressed in his red and purple uniform, he finally walks up to the base itself, wraps up the troops in hardened magnetic waves, and puts a magnetic force field around Cape Citadel.

The army surrounds Cape Citadel but can't break through the magnetic barrier. Magneto isn't prepared to battle mutants, and the X-Men use their special powers to smash through the shield. After several minor skirmishes, Magneto decides the base isn't worth the effort and uses his magnetic repulsion power to fly away. The X-Men have won the first battle, but the war is far from over. Magneto never once mentions the reason he wanted to capture Cape Citadel or what he would have done if he had been able to keep it under his control. Among his worst faults as a supervillain, Magneto rarely plans very far in advance, and he doesn't believe in backup strategies.

Magneto returns a few issues later to battle the X-Men again, but in this adventure he brings friends. Magneto's assistants have no group name, but Marvel likes to refer to them collectively as the Band of Evil Mutants (later known as the Brotherhood of Evil Mutants). Among Magneto's earliest recruits are the Mastermind, Toad, the Scarlet Witch, Quicksilver, and from time to time, the Blob. Though the odds are evened up, the Evil Mutants are never able to beat the X-Men, partly because the good guys fight together as a team while the evil mutants fight as selfish individuals.

In all of the fights that follow for the next few years, Magneto comes across as an arrogant, self-centered madman with ambitions of enslaving all of humankind, including mutants, and ruling the world. While always evoking the power of magnetism, his powers have little actual connection with magnetism or magnetic fields. Tying people up with invisible bands of magnetic form sounds terrific, as does flying around on magnetic force beams, but neither power makes any sense. Fortunately, since no explanation is ever offered for any of Magneto's powers, readers really can't complain that they're impossible.

Magneto makes it clear to anyone who will listen that he's the most powerful mutant in the world. Despite the fact that the X-Men stop his every plan, sooner or later he's going to smash down everyone who stands in his way. Magneto's been imprisoned multiple times (on Earth as well as on other planets), gravely wounded, de-aged to a baby, lost his memory, and supposedly killed (numerous times and in numerous fashions). Yet he never gives up. He's not only the world's most ambitious mutant, he's also one of the most persistent. Magneto believes that mutants should rule the Earth and that he should rule all mutants. And he's willing to do anything at all to make that dream come true.

Smug, arrogant villains lose their appeal after a time. Just as the heroes in comics need to change, evolve, and develop fuller personalities, so do the villains. In Magneto's case, his past is an empty book, and for years nothing of his previous life before he attacked Cape Citadel was known. Then, slowly but surely, the various comic writers in charge of X-Men began to retcon his past.

The term retcon was invented by longtime comic book writer and editor Roy Thomas, and it stands for "retroactive continuity."

Thomas came up with the term in 1983 to describe "changing and adding details of a fictional setting, often without providing an explanation in the context of the setting."[2] Retconning often takes place in comic book series to flesh out and give greater depth to certain characters and their earlier lives. However, retconning is also used in television series and movies.

Perhaps the best-known science fiction television example of retconning took place in the universe of *Star Trek*. In the original series, Klingons looked much like humans except for minor makeup changes. In the later *Star Trek* series, including *The Next Generation* and *Deep Space Nine*, advances in mask-making and makeup gave Klingons a much more alien look. Instead of merely ignoring the differences between Klingons in the first series and the later series, an episode of *Deep Space Nine* involving time travel and tribbles addressed the differences straight on. When Commander Worf and his companions from *Deep Space Nine* observed some Klingons at a bar in the twenty-second century and they didn't have facial ridges, Worf was extremely embarrassed. When asked about the lack of ridges, Worf muttered something like, "It is not something we like to talk about." Though the interchange didn't explain the different looks, it did acknowledge them, thus bridging the gap between the two series and making it clear they all were part of the same universe.

In comic books, retconning usually refers to changing or rearranging the early history or life of a superhero or supervillain. It's usually done to give the hero a more modern approach. It is also used to add bits of history to a character whose past directly influences his present actions. While many superheroes and supervillains have histories that are described in great detail, others have big blanks before their transformations. In the movie version of *The Hulk*, Bruce Banner's childhood and father assumed an importance they never had in the comic book. In the *X-Men* comic, Scott Conners aka Cyclops seemed to be normal until it was revealed years after the series began that he was raised in an orphanage by the diabolical Mr. Sinister and had an estranged brother who also possessed mutant abilities.

Retconning fills in the holes and explains the mysteries. It's also a wonderful device for complicating simple stories and turning ordinary adventures into epics. For Magneto, retconning gave meaning to his life and an explanation for his feelings. Unfortunately, it didn't make his powers any more believable.

Over the years, numerous comic book writers have used Magneto's past as a springboard for new ideas and unusual concepts. Thus, his background is more complex than the backgrounds of most other characters in the X-Men universe. And in more than one or two spots, it contradicts most of what we know about Magneto from his earliest encounters with the X-Men. Which is why retconning is a great tool if used correctly and not so great if used without first investigating what's already known about the character. Because of these multiple expositions of Magneto's past, much of the information about him is suspect. For example, after forty years and hundreds of appearances, his true name is still in doubt. For simplicity's sake, we'll stick to the version of Magneto's history that makes the most sense.

Erik Magnus Lehnsherr was born in the late 1920s in or close to Gdansk, Poland. He was a member of a large Polish Jewish family. Several years after the Germans overran Poland, a group of Nazi soldiers came to his village and gathered all of the Jews and other "undesirables." The prisoners were driven into a nearby forest where they were ordered to dig a large pit. Once the hole was completed, all of the prisoners were lined up at the edge of the pit and shot. Only Erik survived, untouched by any of the bullets. Evidently, his magnetic power repelled all bullets aimed at his body.

He did, however, tumble into the pit. It took him hours to dig himself out, and when free, he was almost immediately recaptured by the Nazis. He was then sent to Auschwitz.

It was in Auschwitz that Erik discovered just how horrible ordinary people can act when dealing with minorities. He realized that those different from the majority were always in danger. This truth remained with him.

In winter 1945, Erik Magnus Lehnsherr saved a young gypsy woman named Magda from being killed by an SS officer. Erik and Magda escaped the camp. After some wandering, they settled in a village in the Carpathian Mountains. Magda gave birth to a daughter who they named Anya. However, Erik wasn't destined for the simple life.

Anxious to get an education, Erik moved his family to the Soviet city of Vinnitsa. It was here that his power over magnetism finally revealed itself. Working as a laborer, Erik was cheated out of his first day's wages by his boss. Angry, he mentally sent a crowbar flying across the lot at the foreman but missed. Fleeing back to the inn where he had left his family, Erik was attacked by soldiers sent by his boss. The inn caught fire. Magda escaped, but four-year-old Anya burned to death before her father's eyes. Using his powers for the first time, Erik killed all those who prevented his daughter's rescue.

Fearful of her husband's mutant powers, Magda fled, never telling him she was pregnant with twins. Erik never saw her again. She died soon after giving birth. The twins, Wanda and Pietro, were raised by a gypsy family and didn't learn they were Erik's children until many years later. Both were mutants. Pietro possessed superhuman feet and called himself Quicksilver, while Wanda controlled probability forces and became known as the Scarlet Witch.

Alone, Erik made his way to Israel, where he served as an orderly at a psychiatric hospital that helped Holocaust survivors. It was here that he met Charles Xavier, the mutant later known as Professor X, the founder of the X-Men. The two brilliant men became friends, though they had strongly opposing views about the future of humanity. Xavier felt that mutants needed to coexist with humankind while Lehnsherr believed cooperation between humans and mutants was impossible.

After thwarting a plan to establish a Fourth Reich, Erik disappeared only to turn up in Brazil hunting down Nazi war criminals for a nameless U.S. government agency. It was here that he started calling himself Magneto. As usual, things didn't proceed well with his new occupation, and Erik's handler, Control, tried to kill him. The attack failed. Erik's lover, a beautiful Brazilian doctor, died. He killed Control, then disappeared.

This time, Lehnsherr remained hidden for more than a decade. When he returned, his magnetic powers were totally under control. It was at this time that the X-Men saga first began, with his encounter with Xavier's team at Cape Citadel. Soon afterward, the Brotherhood of Evil Mutants was introduced. In those early stories, there's no mention of Erik's terrible past or of his relationship with the Scarlet Witch and Quicksilver. Those facts were retconned into Magneto's story.

Magneto established headquarters for himself and his minions on an asteroid hovering over Earth. He dubbed the planetoid Asteroid M. Modesty was not one of Magneto's traits. When the world's governments discovered the location of Asteroid M, they tried to construct a powerful magnetic grid to keep Magneto from returning to Earth. Magneto retaliated by generating an electromagnetic pulse that shut down nearly all electricity on the Earth. When he relented, the United Nations decided to leave Magneto strictly alone.

Many encounters later, Magneto finally realized the futility of battling humanity for control of the world. When Professor X nearly died, it was Magneto who took over Xavier's school and helped educate future X-Men. For a long period, Magneto wavered between war with humanity and compromise. The situation finally ended peacefully with the United Nations giving Magneto the island of Genosha as a safe haven for mutants.

Unfortunately, the mad twin sister of Charles Xavier attacked Genosha with its large mutant population and slaughtered nearly every inhabitant of the island. For months, it was stated that Magneto had died during the attack. A new Marvel editorial policy stated that characters killed in stories would stay dead and not be resurrected by some tricky maneuvers. However, just recently, Magneto made his return to the pages of *X-Men*, proving once again that retconning is still alive and well in comic books.

Magneto calls himself the Master of Magnetism. What exactly does that mean? He also claims to be the most powerful mutant in the world, and in the X-Men universe, that's pretty powerful. Is Magneto all talk or can he back up his claim of near omnipotence? Before we investigate Magneto's powers, let's first study some facts about magnetism and electromagnetism.

There are four basic forces in the universe. Each governs a specific way that particles of matter interact. All other forces that we're aware of, from electrical power to wind power to explosions to muscular power, are based on these four elementary forces. Just as the subatomic particles known as quarks combine to form atomic particles such as electrons, neutrons, and protons, which then combine to form atoms, then molecules, then objects, these four forces combine to form all forces in the universe.

It's fairly obvious that force is exerted by one object on another when the two objects are touching. However, certain forces like gravity or magnetism don't involve any actual physical contact between objects. Instead, these forces are carried from one object to another by special particles such as photons, mesons, and gluons. We've discovered most of these particles, but modern science is still looking for one more that theoretically should exist.

Two of the four basic forces work within the nucleus of atoms and are known as the unfamiliar forces. The other two forces, which we experience in our daily lives, are known as the familiar forces.

The unfamiliar forces are known as the strong (or nuclear) and the weak forces. The strong force is the most powerful one in the universe. It binds together the protons and neutrons that form the nucleus of an atom. The force is carried by particles known as mesons and holds the protons and neutrons together. It only works over a very short range and is not effective beyond 1 fentometer (10^{-15} meters).

However, it's been discovered that the strong force carried by mesons is a by-product of the strong force inside protons and neutrons. Protons and neutrons consist of quarks, and the strong force that holds quarks together is carried by a particle called a gluon, about which very little is known.

The second most powerful force in the universe is electromagnetic force, which only has one-fourteenth the power of the strong force. Electromagnetic force theoretically extends to infinity, but its strength diminishes quickly due to shielding. Electromagnetic force acts between particles carrying an electric charge. Electrons and protons both have the same amount of electric charge. However, electrons have a negative charge, while protons have a positive charge. In an atom with the same number of electrons and protons, the charges balance out and the atom is electrically neutral. If an atom has more electrons than protons, it's said to be negatively charged. If it has more protons than electrons, it's positively charged.

Photons transmit this force between electrons (negative charge) and the nucleus of an atom (positive charge). A stream of photons is called electromagnetic radiation. Cosmic rays, X-rays, light, and microwaves are all examples of electromagnetic radiation.

An electric current is a movement of charge. Magnetism is basically a force between electric currents. Objects with opposite charges attract each other, and objects with similar charges repel each other. Coulomb's law, formulated by French physicist Charles Augustin de Coulomb in the eighteenth century, enables us to calculate the strength of the attraction or repulsion.[3] Magnetism will attract iron, nickel, and cobalt. Combinations of these metals made into alloys can become permanent sources of magnetism—that is, magnets.

Magnets have a north and a south pole, although on a global scale these are misnomers. The north pole is actually seeking the Earth's magnetic north pole, while the south pole of a magnet is seeking the Earth's south pole. In magnetism, like poles repel each other, just as in electricity like charges repel, so the north pole of a magnet is really its south pole.

The third most powerful force is known as the weak force. It's the force responsible for the decay of fundamental particles in atoms. The weak force is one–one millionth weaker than the strong force. The particles that transmit the weak force are called the W and Z particles, and they exist for very short instances.

The least powerful of the four forces is gravity. Compared to the strong force, it is barely noticeable in an atom. Gravity is 10^{-38} of the strong force.[4] However, gravity affects particles over huge distances and is cumulative. It exists anywhere there is matter. While the gravitational force of one atom is small, the combined gravitational forces of many billions of atoms are what keeps us on Earth, and on a much larger scale, are responsible for the Earth rotating around the sun. Gravity is thus the most important force in the universe. Scientists are still searching for some sort of particle that transmits gravity (which they call gravitons) and are also trying to find gravitational waves.

Most scientists believe that these four basic forces are merely variants of a single universal force that appeared at the instant of the big bang, when the universe began. Scientists have been searching for this one universal force and how it unifies all the other forces for many years, and it is one of the great unsolved mysteries of the universe. This unification of forces is known as the grand unified theory. Over the past thirty years, many scientists have come to believe that the string theory—that is, all elementary particles are manifestations of the vibrations of one-dimensional strings—is the key to the grand unified theory. Unfortunately, proving the string theory isn't easy.

Electromagnetism is the second most powerful force in the universe, and Magneto controls magnetism and thus, most likely, electromagnetic force. With those two thoughts in mind, let's take a look at some of Magneto's powers.

Can a person control magnetism just by using his or her mind? As pointed out in our previous book about comics, *The Science of Superheroes*, there's no widely accepted scientific explanation for human extrasensory perception (ESP). There's no thoroughly documented proof that it actually exists. However, there's definitely a link between magnetism and the brain. A human brain has billions of neurons that discharge electricity whenever stimulated. These electrical patterns are called brain waves, which are affected by magnetism in ways that cannot be fully explained.

For example, according to a series of tests performed at the Centre for the Mind in Sydney, Australia, magnetic waves improved the drawing skill level of seventeen volunteer children in less than fifteen minutes. Professor Allan Snyder, who runs the Centre for the Mind, uses a Medtronic machine to supply electromagnetic pulses that he claims can give extraordinary skills to ordinary people. Snyder believes that people gain "isolated pockets of geniuslike mental ability" after exposure to transcranial magnetic stimulation.[5]

Many scientists have linked extremely small magnetic changes in the brain to deep depression. A few researchers, mostly in England, are predicting that within a few years routine magnetic brain scans will be able to predict the beginning of Parkinson's disease or severe depression. Like most scientists, they know that the brain, human thoughts, electricity, and magnetism are very closely related. None of them are willing to state that some sort of mutation in the brain would allow a human to control magnetism. But very few are willing to deny that such a mutation is impossible. Magneto might never exist, or he might be walking among us right now. Not even the greatest doctors and scientists in the world can say for sure.

At the conclusion of his first battle with the X-Men in September 1963, Magneto escapes by flying away via magnetic repulsion. It's a transportation method he uses quite often in the stories that follow. Similarly, in the first *X-Men* movie, Magneto floats from place to place, not flying but seemingly levitating across the ground and up the side of the Statue of Liberty.

As discussed earlier, all magnets have north and south poles. Two of the same poles repel, while two opposite poles attract. Using this basic truth about magnets and magnetized objects, scientists and salespeople have sold levitation devices for years. A sample of many such constructions can be found by doing an Internet search for "magnetic levitation."

In simplest terms, a magnetized metal ball is placed between two stationary magnets: one above the ball and one below. There's a thin layer of air between the magnets and the ball that barely matters. The magnet above repels the ball downward, while the magnet below repels the ball upward. The lower magnet is stronger than the upper one, since the force pushing the ball upward has to counter the magnetic force pushing downward as well as the gravitational pull of the Earth.

Acted upon by the repulsive forces from above and below, the ball hovers at an equilibrium point where the two forces cancel each other out. In theory it sounds perfect, but in real life it doesn't work. That's because of a theorem proved by Samuel Earnshaw in the nineteenth century that showed it is impossible to achieve static—that is, stable—levitation using any combination of fixed magnets and electric charges. Fortunately, for Magneto and scientists around the world, experimenters soon found ways to get around Earnshaw's theorem. The easiest and most common method of avoiding the consequences was to use diamagnetic objects instead of a magnetized metal ball.

Diamagnetism was discovered by the famous scientist Michael Faraday in 1846, but it was largely ignored as no one had any use for it. Ferromagnetic material such as iron is strongly attracted to both poles of a magnet. Paramagnetic material such as aluminum is weakly attracted to both poles. But diamagnetic material (and there are many types) is weakly repelled from both poles of a magnet. In fact, the magnetic force for diamagnetic material is a million times less than the magnetic force of magnetic material.

Electrons in atoms of iron, cobalt, and nickel align with electrons in nearby atoms, forming regions called domains that possess a strong magnetic field and thus form ferromagnetic material. Atoms that have a single, unpaired electron are paramagnetic. The electrons in such material are weakly attracted to magnetic poles. Paramagnetic material includes hydrogen and lithium. Atoms in which all of the electrons are paired with electrons of opposite spin, so the orbital currents equal zero, are diamagnetic. The magnetic fields associated with this material are repelled by other magnetic fields. Among the many diamagnetic substances are helium, bismuth, and water. Most living things are diamagnetic, but the force is so small that it's barely ever noticed.

Diamagnetic materials don't obey Earnshaw's theorem. That's because diamagnetism involves electron motion around the nuclei of atoms, which according to quantum theory can't involve fixed magnets and electric charges. In 1991, Eric Beaugnon and Robert Tournier performed a series of experiments working with strong magnets, a magnetic field, and a drop of water (which is diamagnetic). They discovered that a magnetic field of 10 tesla, which is only about seven times as powerful as most permanent magnets (which have a field strength of 1.5 tesla), or one hundred times the strength of a refrigerator magnet, was capable of levitating the water drop. In other laboratories throughout Europe, scientists performing the same experiment levitated liquid hydrogen, liquid helium, and frog's eggs. Probably the most exciting moment came at High Field Magnet Laboratory, University of Nijmegen in the Netherlands, where a living frog and a living mouse were levitated into the air for an hour or more. Afterward, the animals were closely examined. They exhibited no ill effects from their levitating experience.[6]

While further experiments in diamagnetic levitation have yet to lift a human being off the ground or allow a levitating object to move from place to place, there's little doubt that such results will be achieved as we learn more about electromagnetism and wave motion. Since Magneto controls incredibly powerful magnetic forces, it seems fairly safe to say that his ability to levitate his own body isn't a surprise.

Magneto's main goal in life is to create a society where the minority mutant population of Earth can live in peace and not worry about being persecuted by the majority of ordinary humans. In pursuit of this noble dream, he's fought many battles against human foes, following the old adage that the end justifies the means. His most effective means of attack—the one that won him an island empire where all mutants could live in peace—was disrupting the Earth's magnetosphere. But creating this worldwide disaster almost killed Magneto and imperiled the safety of all life on Earth.

There is another, easier method Magneto could use to threaten the stability of the world without menacing the existence of the Earth. It's a power he controls and has used in limited capacity before. It's the awe-inspiring force of the electromagnetic pulse (EMP) bomb. It's a weapon that every nation in the world is trying to master at this very moment.

Forty years ago, testing nuclear weapons in the atmosphere was one method of gauging their destructive powers. As far back as the 1940s, some scientists theorized that such atmospheric blasts would create gigantic electromagnetic pulses, but no one knew for sure if this idea actually made sense until 1962. That's when the U.S. government set off a high-altitude nuclear explosion called Starfish Prime approximately 800 miles from Hawaii. The explosion caused a massive EMP that wreaked havoc with electrical equipment throughout Hawaii as well as disrupting radio communications.

The science behind an EMP is fairly simple. An atomic explosion creates a huge shock wave of highly energized photons that collide with oxygen and nitrogen atoms in the atmosphere. These atoms eject a vast wave of electrons speeding outward from the explosion site, an action known as the Compton effect. This massive EMP diminishes over distance, but it is capable of disrupting any electrical device within several hundred miles of the blast.

The strength of an EMP is directly related to how high in the atmosphere it is released. The electric field created by the electromagnetic pulse only lasts for a few seconds, but that's enough time to cause tremendous amounts of damage. Scientists have estimated that an EMP wave set off 200 miles above Kansas could disrupt electrical systems throughout the entire United States. An EMP would cause major damage to electrical systems as well as destroy all types of microcircuitry.

The incredible effects of an EMP wave have been demonstrated in several action movies including *Ocean's Eleven* and *The Core*. Numerous conferences and Senate hearings have been held in Washington to discuss the dangers of an EMP attack. According to expert testimony of retired generals and military leaders, EMP weapons are the next big thing in weapons of mass destruction. Some experts claim that they can be built for as little as $400 and that the plans for EMP bombs are available free on the Internet. These same experts claim that EMP weapons would be the size of a small suitcase. What they do not explain is exactly how these weapons would work without a nuclear bomb acting as the trigger.

Despite all the hype and talk about EMP bombs and EMP weapons, there's been no actual demonstration of an EMP wave being created without an atomic explosion. All of the dire warnings from defense contractors, Internet diagrams, and $400 bomb kits have proven empty. Electromagnetic pulse weapons may be the weapons of the future, but so far they're only seen in comic books. They are the stuff of Magneto. Hopefully, they'll remain that way.

11

Immortality for Some
Vandal Savage and Apocalypse

"In this world," wrote Benjamin Franklin, "nothing is certain but death and taxes."[1] If Franklin were alive today, he might be surprised how many of the world's wealthiest individuals have managed to avoid the second, but he'd be right in assuming that none had escaped the first. Though the length of our lives might astonish him. Or might not, since Franklin lived to be eighty-four. Still, the average life span of someone born in the eighteenth century (Franklin was born in 1706) was approximately twenty-seven years. By the middle of the twentieth century, it had increased to forty-six years. Now the world average is sixty-six.

Of course, that number is deceptive. In East Africa, the average life span is forty-five years. In the United States, men live to an average age of seventy-eight and women to over eighty. These are fairly typical statistics for people living in highly developed industrial societies. Still, while it can be postponed for decades, death is certain. As far as we can determine, there are no immortals living among us in the real world. But what about the unreal world—the world of comic books?

Comic superheroes and supervillains have exhibited all sorts of wondrous powers over the years. Some can fly into space, others can shrink to the size of an atom, and still others can run nearly as fast as the speed of light. They've taken potions, been bombarded by cosmic rays, and been dusted with radioactivity. Did any of these occurrences make them immortal? Are any of them ageless?

In the DC universe, most superhero characters rarely show any signs of getting older, but that's entirely a result of focusing only on adventures taking place when they are young adults or teenagers. However, it's clear that heroes do age off screen, since Superboy grew to be Superman, and the entire Justice Society of America consists of older, if not wiser, superheroes.

In the Marvel universe, the clock is ticking, but much more slowly than in the real world. Spider-Man has progressed from a nerdy teenager to a less nerdy college student to a dedicated inner city schoolteacher. The Hulk has undergone so many life-changing events that he should be one hundred fifty, and the X-Men have had enough adventures to fill several lifetimes. Yet, while Cyclops and Jean Gray and Beast have all gotten older, they haven't gotten much older. Ten years of comic book adventures work out to approximately one year of Marvel comic book time.

The only superhero among the major science fiction comic book characters published by Marvel and DC who might be immortal (excluding magical characters and minor deities like Thor and Hercules) is a mutant who refuses to die. He's Wolverine, one of the X-Men. Wolverine possesses incredible healing factors that work at miraculous speed. No amount of physical punishment, gunfire, stabbing, burning, and worse can harm Wolverine, who is a gifted killing machine. Though he seems to be aging, albeit very slowly, he's nearly two hundred years old and shows no sign of slowing down. Whether his immune system protects him from growing older or he possesses some other mutant power isn't known. As is so often the case in comic books, no one (including the government, scientists, doctors, etc.) seems very interested in investigating Wolverine's genes to see if his powers could be copied to benefit humankind. Instead, everyone thinks of him only in terms of how many people he can kill. Wolverine's situation is comparable to using a cancer cure to get rid of the hiccups.

Superheroes aren't immortal for the same reason they're rarely invulnerable. Even Superman's affected by kryptonite. There's no dramatic tension, no real excitement reading a story where you know that the hero can't be killed. Comic book drama relies on the main characters being in deadly peril. Without the unspoken but ever-present threat of death, comics would be little more than brain teasers with characters.

Ambitious beyond reason, to supervillains living forever means multiple chances to conquer the world. Lose a round to some chance of fate or the intervention of a pesky superhero, it doesn't matter. Super-villains merely retreat to one of numerous safe havens prepared well in advance and live to scheme for another day. Immortality offers villains a chance to rise from menace to mastermind. Undying villains are mem-orable because they can take the long view—the very long view.

Such was the case of Vandal Savage, one of the more memorable villains of the DC universe. Savage was created by Martin Nodell, the creator and artist of the Golden Age Green Lantern, and science fic-tion author, Alfred Bester, who worked as a comic book scripter for DC in the 1940s. His first appearance was in *Green Lantern* #10 (win-ter 1943), but he also fought the original Justice Society of America several times.

Savage returned in the Silver Age *Flash* in 1960. Afterward, he fought numerous battles against the revived Justice Society as well as other DC superheroes. In all his appearances, his aim remained the same: to conquer and rule the Earth. As his name implied (Golden Age comics were rarely subtle), he was a vandal and a savage.

Savage originally was a Cro-Magnon man named Vandar Adg roaming Earth fifty thousand years ago.[2] One day, a meteor crashed to Earth near Adg, and he fell into a deep sleep/coma. When the caveman awoke, he discovered that the meteor's radiation had made him immor-tal. Changing his name to Vandal Savage, the transformed Cro-Magnon man set out to gain control over all humanity. Among his many accomplishments, Savage formed the infamous secret society, the Illuminati, during the last days of Atlantis. He also developed a number of magical powers including superstrength that required blood sacrifices. These powers vanished over the course of several thousand years. Savage was a pharaoh in Egypt and a king in ancient Sumer. He also claimed to have been both Julius Caesar and Genghis Khan.

Like most immortals in science fiction and fantasy novels, Savage came to realize that Shakespeare knew what he was writing when he said, "Uneasy lies the head that wears a crown," and he decided it was a lot safer being the power behind the throne than actually sitting on it. According to Savage, he served as advisor to William the Conqueror, Napoleon, and Otto von Bismarck.[3]

In World War II, Savage was one of a number of comic book masterminds behind the Axis powers. It was during this period that he tried to worm his way into the U.S. government but was thwarted by the Green Lantern. Later in the war, he formed the Injustice Society in an attempt to destroy the Justice Society of America. His plan failed, and Savage disappeared for decades.

Now, you might wonder why Savage didn't merely decide to wait out his enemies. One of the greatest benefits of immortality is that you can live in a relaxed manner, and given enough time, everyone who bothered you will die off. In 1945, Savage could have bought a nice villa in the Brazilian rainforest and raised cocoa for a hundred years. Why stir up trouble when time heals all wounds? But, since this is *The Science of Supervillains* and not *The Psychology of Supervillains*, we'll leave it to some other ambitious writing team to explain supervillain megalomania.

Savage returned in the early 1960s to capture the retired members of the Justice Society of America. The new Flash and the old Flash teamed up to defeat him. Because of Savage's attack, the Justice Society decided they needed to return to action.

More recently, Savage regained his faded mystic powers and became a Manhattan drug lord. Needing help, he took control of the Illuminati. In his unending quest to rule the world, he's battled everyone from the Justice Society of America to the Teen Titans. Like most men fifty thousand years old, he's used to having his own way, and he doesn't believe in compromise.

Marvel's most notable immortal villain is named En Sabah Nur ("The First One"), but he prefers to go by the name Apocalypse. He's even less subtle than Vandal Savage.

According to Apocalypse's origin story, modestly titled "The Rise of Apocalypse," baby En is found in 2967 B.C. in the Sahara Desert by Baal, the leader of a band of thieves and murderers. Baal assumes, as do readers, that the baby is so ugly he has been left there to die by the people of Akkaba. Considering he has gray skin and blue-gray hair, the assumption seems fair, though why he is just left to die instead of being killed outright is never fully explained. It's one of those fascinating loose ends that comic book writers like to insert in their stories to be picked up years later and turned into a new story, but in this case it seems to be totally forgotten. We never learn who Apocalypse's true parents are, or even if he really was abandoned by the Akkabas. It doesn't matter much. What is important is the fact that Apocalypse was the first—the very first!—mutant on Earth.

Baal senses that the child is somebody important, so he raises En as his own son. Soon we learn that the pharaoh Rama-Tut is actually a visitor from the future who has returned in time to find this legendary first mutant and use him to conquer the future. After a bunch of melodramatic happenings including Baal being killed, En stumbling across some of Rama-Tut's futuristic technology, an ancient prophecy, En falling in love but being rejected by a princess as being too ugly, and more, En Sabah Nur finally develops his powers and causes havoc throughout Rama-Tut's palace. Rama-Tut escapes back to the future, where he later becomes another Marvel villain, Kang the Conqueror. En wanders into the desert and disappears for approximately four thousand years. We later learn that during this time he's been worshipped as a god in Persia, Mexico, India, and, of course, Egypt.

When next seen in the flesh (*X-Force* #37), En is in Mongolia in the twelfth century. He enters an abandoned alien spacecraft. When he comes out, he's transformed into Apocalypse. In his final mutant form, En possesses complete control over his molecular structure and can alter his body into any shape or form he desires. He's also developed a very Darwinian philosophy: survival of the fittest. Apocalypse feels that mutants should rule Earth, but only those mutants who prove themselves strong. So he tests likely candidates to make sure they're tough enough to meet his standards.

As Apocalypse, En's powers are so great that they rage through his body like a fever. He needs more rest, so he undergoes periods of cybernetic sleep and regeneration. Awakening in London in 1859, he recruits a mad genetic researcher, Nathaniel Essex, who he transforms into the aptly named Mr. Sinister. Unfortunately, Sinister turns out to have his own agenda, and he and Apocalypse part on less than friendly terms.

Apocalypse awakens again in the twentieth century when Nathan Summers, the son of Scott Summers and Madelyne Pryor, is born. The child is so powerful a mutant that Apocalypse is determined to kidnap him and use him for his own unstated purpose. That doesn't happen, but Apocalypse does manage to infect the baby with a techno-organic virus before making his escape.

The purpose of the virus is to convert man (or in this case, baby) into machine. There's no cure for it on present-day Earth, but a time traveler from two thousand years in the future arrives to take the baby into the future where a cure supposedly waits. Though Scott is reluctant to give up his son, he does what he must to ensure the child's survival.

That's when the plot shifts into high gear. It turns out that the visitor came from a future ruled by Apocalypse and a cadre of evil but very powerful mutants. The time traveler is part of an underground movement called the Askani, which is headed by Scott and Jean Gray's daughter, Rachel, from another dimension. There's no cure for baby Nathan, but the virus can be stopped from spreading any further. Apocalypse's men come hunting for the baby, but the Askani have cleverly created a clone of Nathan, and the clone is captured. Nathan, who is partly living metal from the techno-virus, is rescued by Scott and Jean, whose minds have been transferred into future bodies by daughter Rachel. On the run for the next ten years, Jean and Scott raise Nathan.

Meanwhile, apocalypse raises the clone created by the Askani, thinking the child is Nathan. He names the boy Stryfe. En's powers have grown so strong that he's been forced to change bodies through mind transference every few years. He believes that Stryfe will serve as his perfect body. He never gets a chance to find out.

Nathan, at eleven, gets trapped in Apocalypse's fortress, and through a combination of his mutant powers and techno-organic skills, destroys Apocalypse. Stryfe thus becomes the ruler of the future and Nathan's major new foe. Years later, an adult Stryfe travels back to modern-day Earth to try to gain control of the past. Nathan follows his clone back to the past and takes on a new identity, Cable. Not only does Cable finally defeat Stryfe in the past but he manages to destroy Apocalypse as well. Thus, he wipes out his own time line, which depended on Apocalypse being alive thousands of years in the future.[4]

Apocalypse, if you're keeping score, was killed twice by Nathan (Cable): once in the future, once in the present. Which proves that immortality doesn't guarantee invulnerability. Still, Apocalypse lived for somewhere around eight thousand years, and Vandal Savage was born fifty-two thousand years ago. Neither man conquered the world, but if they'd invested their money properly in a good bank, they could have bought it.

Is there a chance that humans will someday live for hundreds or even thousands of years? Is immortality possible, or is it an idea only for comic books? The answers might surprise you.

Recently, two scientists researching the aging process made the news by placing a bet on whose descendant would live the longest. S. Jay Olshansky of the University of Illinois at Chicago School of Public Health and Steven Austad of the University of Idaho Biological Science Department each contributed $150 to a trust fund that will pay out $500 million (with the help of more contributions over the years and interest) to the oldest child in the two families in 2150. Austad believes that people will live with full memory and motor skills to be well over one hundred fifty while Olshansky believes that one hundred thirty is the upper limit.

Austad thinks Olshansky "has less faith in the rate of progress in medicine than I, a practicing biomedical researcher, have. I also think we are closing in on the fundamental process of aging."[5] Olshansky disagrees with Austad, saying, "The only way we can push an individual out for an additional 28 years beyond the world's longevity record of 122 years . . . is if we alter the basic biology of aging itself."

Still, both men are upbeat on extending the human life span. "The prospects of dramatically increasing human longevity are excellent," says Austad. Olshansky agrees, explaining, "We see ourselves on the cusp of the second longevity revolution. Scientists are on the verge of discovering major secrets of aging."[6]

If there are no heirs alive in 2150 (both men are married with young children), the winner will be determined by a panel of three scientists appointed by the American Association for the Advancement of Science. If Austad's claim is true, the money will go to UCLA; if Olshansky is right, the money will be awarded to Michigan State and the University of Chicago.

Is there a certain age limit beyond which humans can't survive? Not really. Scientists have been predicting how old people can live with a notable lack of success. In 1928, census expert Louis Dublin predicted that the average life span in the United States would never exceed sixty-five.[7]

Most researchers agree that aging is due to cell damage by free radicals, which are atoms or molecules with at least one unpaired electron that causes the atom or molecule to be chemically reactive. A large number of free radicals are created in cells when they produce energy. The free radicals damage a cell's DNA, repair mechanisms, and protein synthesis.

Every time a cell divides, it copies all the DNA pairs that make up its genome. However, sometimes mistakes are made due to the effects of free radicals. These mistakes are passed on through the DNA, resulting in defective cells. The bad genes produce distorted proteins, causing molecular damage to the body. This damage progresses as the human body ages.

Free radicals are also involved in inflammation, which occurs when the immune system attacks bacteria and viruses. The immune cells produce vast numbers of free radicals to rip these diseases apart. Unfortunately, sometimes when the bacteria (or virus) are destroyed, immune cells continue to make huge numbers of free radicals. Such inflammation has been closely linked to diseases associated with aging such as cancer, arthritis, and Alzheimer's disease.[8]

Scientists are fairly certain that if the manufacture and intervention of free radicals could be slowed, aging would also slacken and people would live longer. So far, the only method discovered to cut down the number of free radicals is to cut the amount of food you eat by two-thirds. Restricting calories to a near starvation level is the only agreed-upon technique at present that will extend human life.

Happily, not all scientists are convinced that starvation diets are the answer to immortality. Regenerative medicine is making huge strides in curing many of the diseases that cut life short including Alzheimer's disease, strokes, paralysis, heart disease, and cancer. The magic bullets making these cures possible are stem cells, which are mentioned in chapter 7.

Understanding the power of stem cells is as easy as planting a garden. Cut off a piece of your favorite plant and put it into some potting soil. Within a few weeks, you have a new plant. Plants maintain small amounts of stem cells in their tissues. These cells have the ability to reproduce and change into all the specialized cells needed for the new plant.

In the human body, stem cells exist in early human embryos and in adult bone marrow, brains, and livers. Embryonic stem cells are the most useful as they seem to be able to produce all the specialized cells of the human body. Adult stem cells are much more limited in the type of cells they can reproduce. Unfortunately, embryonic stem cells must be taken from five-day-old human embryos, which are destroyed in the process, raising powerful issues about ethics.

Neural stem cells injected into paralyzed rats have actually traveled to damaged areas of the animal's spinal cord and repaired the damaged nerves, enabling the rats to walk. Stem cell research into Parkinson's disease has shown amazing progress. Strokes and Alzheimer's disease are two other incurable medical problems that stem cell medicine may someday cure.

Cancer is another area where stem cell research has made huge strides. It occurs when cells start growing uncontrollably. Healthy cells have a self-destruct program built into their structures, but in cancer cells that self-destruct mechanism doesn't work. By studying stem cells, scientists have found a chemical signal that causes some cancer cells to destroy themselves. Gilvec, a new drug based on stem cell research, sends that signal to cancer cells. Gilvec is the first of a new type of cancer remedy that might one day stop this disease.

Curing diseases like cancer and Alzheimer's will definitely extend human life. But not all medical problems can be solved by stem cell cures. Actual immortality needs something more.

In Russia, Lyudmila Obukhova of the Institute of Biochemical Physics of the Russian Academy of Sciences has a very different approach on how to achieve immortality. Obukhova is researching a scientific technique known as *crionika*, where lowering human body temperature extends human life span. According to Obukhova, a drop of several degrees in human body temperature results in a life span of 120 to 150 years. A drop of two more degrees, if possible, would lead to people living to 700 or greater.[9]

Cryotherapy was originally developed in Japan in 1975 by scientist Tosimo Yamauchi. He used cold air at 100–180 degrees below zero (Celsius) to treat the joints of patients suffering from rheumatoid arthritis. The results were so dramatic that soon other doctors were using Yamauchi's criosauna to treat skin diseases, excessive weight, and problems with the immune system.

In the United States, where crionika is better known under the name cryonics, the concept of freezing people to make them immortal has been the subject of science fiction novels for the past hundred years. Now it seems closer to reality. In cryonics, people with incurable diseases are frozen when they die, placed in a container of liquid nitrogen, then revived in a future where science has discovered a cure for their disease as well as a foolproof method of reviving them. The booming field of nanotechnology, as discussed in chapter 6, is seen as the key to cryonics, with millions of atom-size machines traveling through a person's bloodstream, repairing all sorts of damages at the molecular level, including the effects from being frozen.

Russian scientist Mikhail Solovyov sees cryonics as the doorway to an immortal society. Solovyov believes that special communities will be centered around a frozen nitrogen-producing plant. Death will be nearly abolished as people will routinely be frozen and years later returned to life and cured of disease. Solovyov believes such communities will spring up in the next twenty years or less.

The biggest problem is that no animal greater than microscopic size has ever been frozen and successfully revived. Water in cells expands when frozen, which disrupts the cell. It's a problem that hasn't been solved, though a preserving technique called vitrification seems not to bother the cells as much. Whether it works isn't yet known, since no one cryonically frozen has been revived as of this date.

Cryonics sounds like one of Apocalypse's techniques to remain young and powerful. It's a science that's been ridiculed by comedians and cartoon shows for the past thirty years. Yet in the end, it's the cryonicists who might have the last laugh.

12

Have Surfboard, Will Travel
The Silver Surfer

Great heroes need great villains. Without an immediate and believable threat to the lead character, a story loses suspense and excitement. The more powerful the hero, the more powerful the villain must be. That's a problem that DC Comics writers discovered about Superman fairly early in his career and have been trying to adjust ever since. It's the same problem every superhero comics company faces sooner or later. How can your characters demonstrate incredible heroism if they're battling ordinary bad guys robbing grocery stores and gas stations?

The answer, of course, is to make sure that there's a handy supply of costumed supervillains around to meet any emergency. Most of the time, in comics this means making sure there are enough stellar and interstellar, higher-dimensional and interdimensional, invulnerable and indestructible evildoers scattered about the universe so that there's always room for another invasion force to make trouble. In Marvel Comics, incredibly destructive troublemakers are commonplace. If anything, sometimes they're much too powerful for their own good. Defeating such characters usually means throwing logic out the window. Consider, for example, the cosmic being known as Galactus, the Devourer of Worlds.

The name of the story is "The Coming of Galactus!" and it appears in *Fantastic Four* #48 (March 1966). As was the style of the time, the first seven pages of the comic finish off the story line from the month before. The incredible Inhumans are trapped inside an impenetrable dome, separating Johnny the Human Torch Storm from his true love, Crystal, forever. Or at least for a few months so that the Torch can sulk and act like a lonely teenager. Meanwhile, we are told in Stan Lee's melodramatic fashion as the Fantastic Four's chartered jet flies away, "Life goes on," and somewhere in the deep vastness of outer space an incredible figure hurtles through the cosmos—"a being whom we shall call the Silver Surfer for want of a better name!"[1]

The Silver Surfer looks very much like the statue given for the Oscars but is covered in silver paint instead of gold. He guides his interstellar surfboard with movements of his arms, and sways out of the reach of meteor storms and asteroids. He zooms through outer space like a living comet, we're informed, with the wild abandon of the wind itself! Mentioning the wind in the same sentence with comets seems odd, but more outlandish comparisons will be made.

On the next page of the story, we learn that the Surfer rockets from planet to planet, with entire galaxies as his ports of call and the universe itself as his highway.[2] The scene switches from the Surfer dodging meteors to the home planet of the Skrulls, an enemy fought several times by the Fantastic Four, located somewhere in the fifth quadrant of the Andromeda galaxy. A Skrull scientist peers into what resembles the bottom half of a submarine periscope but evidently is connected to a gigantic telescope. The astronomer spots the Surfer sailing past his solar system. Immediately, the chief Skrull orders all lights on all planets in the system turned off. Otherwise, warns the chief Skrull, the Silver Surfer might notice their home world and summon Galactus.

By now, we're aware that we're in for a bumpy ride as far as logic is concerned. How do you black out an entire solar system? Is it possible for civilizations on inhabited planets to suddenly pretend that their world is barren? Can the Silver Surfer detect life only if he sees lights?

The scene shifts quickly back to Earth, which is approximately two million light-years from the Andromeda galaxy. We know from

The Science of Superheroes that light travels approximately 300,000 kilometers per second. Doing some slight calculating, we find that a light-year (the distance traveled by light in a year) equals 9.46 trillion kilometers. Which, again doing some simple multiplication, puts Earth 21 quintillion kilometers from the Andromeda galaxy (which is the nearest galaxy to our own Milky Way). That's 21 with eighteen zeroes after it. An ordinary spaceship traveling very near the speed of light would take two million years to reach Andromeda. Even if we borrowed a starship from *Star Trek*, traveling at 100 times the speed of light, it would still take the ship 20,000 years at top speed to reach the next galaxy. So, just to make things interesting, the Surfer is separated from the entire Milky Way galaxy by 21 quintillion kilometers. And, when he does arrive here, considering that the Milky Way has more than 200 billion stars, that's a lot of suns to search.

Someone on Earth is playing games with our atmosphere, first lighting it up with flames, then filling it with giant meteors—that is, blocking off a view of the surface so that our advanced civilization can't be seen from outer space. The Fantastic Four, of course, are blamed for the trouble. (In comics, crowds turn on superheroes at the blink of an eye). Finally, after lots of melodrama and nothing important happening, we learn that the Watcher, a member of an ancient alien race whose duty is to study Earth and never interfere, has interfered.[3] The Watcher has been working feverishly to hide the Earth from the eyes of the Silver Surfer.

The Surfer, who was just in the Andromeda galaxy, hitches a ride on the energy-crest shock waves of a supernova (which cannot travel faster than the speed of light, but that fact is conveniently ignored) that travels across quintillions of kilometers of space and enters the Milky Way. Somehow, despite the hundred million stars and their possible solar systems that the Surfer passes during his trip, none of them satisfy him until he finds his way to Earth, a small planet circling a G-type sun. The Surfer's not fooled by the Watcher's tricks, and after landing on a Manhattan skyscraper, he sends a signal into hyperspace to summon his master, Galactus, from the far reaches of the universe.

An instant later, the Surfer's clobbered by the Thing and is sent hurtling to the streets far below. But that punch is too late to do any good. Galactus's giant ship materializes over Manhattan, and its huge pilot emerges. Looking at people as if they were bugs, Galactus declares, "This planet shall sustain me until it has been drained of all elemental life! *So speaks Galactus!*"[4]

In the second installment of the Galactus saga, "If This Be Doomsday" (*Fantastic Four* #49), Galactus begins setting up his machinery to drain all the energy first from the seas, then from the land, leaving Earth an empty husk. Galactus has no pity for the inhabitants of Earth, considering them little more than ants. It's much more important that he get the necessary energy, as his existence is important to the universe, though that importance is not spelled out in this adventure or any of those that follow over the next forty years. When we do learn the reason for Galactus to exist, in a 2000 series titled *Galactus the Devourer*, it's not worth the big buildup.

Meanwhile, Ben Grimm's blind girlfriend, Alicia, has found the battered and beaten Surfer and has nursed him back to health. In her conversations and through her gentle, sweet nature, she convinces the Surfer that the people of Earth are worth saving. The Surfer turns from a villain into a hero. At the end of the issue, the Surfer returns to the middle of Manhattan, ready to stand up against his longtime master.

Issue #50 of *Fantastic Four* (May 1966) promised a mind-staggering conclusion to the Galactus saga. The excitement was only slightly muffled by a large insert at the bottom of the cover mentioning that the issue would also cover the Human Torch's first day in college. Stories were told in three book arcs, with the beginning of a new story starting in the same issue with the end of the previous story. It was an old magazine trick applied to comics to keep people reading. It worked as long as the quality of stories was consistent.

In "The Startling Saga of the Silver Surfer," the former Herald of Galactus informs his master that he no longer can help him and must oppose his attempts to destroy Earth. The two fight, using energy drawn from space itself as their weapons. Galactus is the stronger of the two, but he hopes to convince the Surfer to stop fighting and continue as his herald, finding him new worlds to devour. But the Surfer will not agree unless Galactus pledges not to engulf planets with living beings on them.

While this cosmic battle continues above the streets of Manhattan, Johnny Storm returns from another dimension where he had gone on a quest for the Watcher. After navigating through indescribable worlds and bizarre realities, the Torch has obtained the most powerful weapon in the universe, the Ultimate Nullifier. The device looks like a medium-size electric can opener, but evidently, if used precisely right, it can destroy entire solar systems, galaxies, and even Galactus.

When faced by the power of the Nullifier, Galactus agrees to leave Earth. Annoyed with the Surfer's revolt, Galactus condemns the Surfer to remain on Terra by removing his powers over space and time. He then vanishes, but not before uttering the usual soliloquy that seems second nature to bad (but not evil) supervillains:

> At last I perceive the glint of glory with the race of man!
> Be ever worthy of that glory, humans . . .
> Be ever mindful of your promise of greatness!
> For it shall one day lift you to beyond the stars . . .
> Or bury you within the ruins of war!
> The choice is yours![5]

In three quick issues, the Silver Surfer went from a cold, unfeeling, space-surfing minion of the greatest destroyer in the galaxy to a kind, noble, understanding hero. It was one of the fastest changes in Marvel history, and the fans loved it. The Surfer guest starred in a number of other issues of *Fantastic Four* before getting his own comic book series that ran for over one hundred fifty issues. Repeatedly, he found himself battling Galactus and new heralds recruited by the World Destroyer. Though Galactus finally died in *Galactus the Devourer*, it's always possible he might rise again through some unique tampering with time, space, and reality. As of this writing, the Silver Surfer is once again scheduled to appear in a new monthly Marvel series.

The first meeting of the Fantastic Four and Galactus was considered by many comic book fans one of the highlights of the Silver Age of Comics. The story was a grand space opera very much in the style of the early science fiction pulp magazines. The Fantastic Four battled with all their might against a villain who wasn't actually evil, just so powerful that he considered humanity little more than annoying insects. The story stretched the limits of the Marvel universe. While the Fantastic Four were the mightiest team of heroes on Earth, their brush with Galactus made it clear that the universe around them was a much tougher place than anyone had imagined.

Logic, which is often ignored when dealing with big menaces from outer space, gets trampled in the Galactus story. It's never made clear why Earth was such a desirable planet for Galactus's lunch. Nor is it ever explained exactly how the Silver Surfer found Earth among the presumed millions of planets in our galaxy. Exactly what energy Galactus plans to drain from the planet is never adequately addressed. On a more mundane level, it's never mentioned what Galactus does when he's not devouring planets.

Some of these puzzles are answered in the pages of the *Silver Surfer* comic book, others in the various Galactus guest appearances in other Marvel comics. The origin of Galactus is told in *Thor* #169 and is reprinted in *Super-Villain Classics* #1 (May 1983). It turns out that Galactus is the only survivor of the universe before ours and is thus fifteen billion years old. Which works out to a lot of planets being devoured during the course of his life.

We learn that the Silver Surfer is actually neither robot nor cyborg but an alien named Norrin Radd of the planet Zenn-La. When Galactus threatened to devour Zenn-La, Radd saved the planet by volunteering to be Galactus's herald, seeking out new worlds for the planet killer to devour. Radd, transformed into the Silver Surfer, was only supposed to guide Galactus to uninhabited planets. But somehow he forgot his plans and led Galactus to Earth.

Interesting stuff, but what do Galactus and the Silver Surfer have to do with modern science? Oddly enough, when the Surfer was first introduced into the Marvel universe, his scientific connection hardly existed. Yet over the years, as we've learned more about our solar system and our sun, the Surfer's method of travel has undergone a change from impossible to unlikely to plausible. Sometimes, a wild guess does come true.

Nearly four hundred years ago, astronomer Johannes Kepler noticed through his telescope that the comet tails seemed to be blown about by what he assumed was a solar wind. Feeling that this motion was proof that winds, which he called "heavenly breezes," existed in outer space, Kepler concluded that ships might be able to fly through space using sails to catch these breezes.

However, over the course of centuries, we've come to realize that space is a vacuum with neither air nor wind. Kepler saw the pressure of photons in space on dust particles released by the comet as it is moving. This photonic pressure is a very light force that isn't visible on Earth because the frictional forces in our atmosphere are so much greater. Therefore, it is only usable in the vacuum of outer space.

This photonic pressure in space is the result of photons and particles that radiate from the sun, causing what scientists call solar wind. This solar wind is a very light force, but scientists are anxious to find a way to harness it because it would be a fuel that wouldn't need to be carried as part of a rocket's payload. Chemically powered rockets use fuel in short powerful bursts, then coast to their destinations. If spaceships could use solar power, they could continually apply force to the rocket. With continuous power, a rocket could make course corrections more often and thus be steered much more accurately. And over a long period of time, the amount of force applied to the sail of the rocket would far exceed that which could be gotten from any chemical fuel.

NASA is working on a program to develop solar sail technology. It's felt by many scientists that solar sailing would be extremely useful on a number of future missions. Solar sails are comparatively inexpensive to make and are able to deliver high-performance propulsion.[6]

The actual sails are made of large, flat, smooth sheets of film supported by ultra-lightweight frames. These sails must be incredibly thin and lightweight to be easily transported into space. A special material has been developed to make the sails. It has a carbon composite that was developed by the Energy Science Laboratory in San Diego, California. The compound is strong enough to resist the intense heat of the radiation in space (up to 2,500 Kelvin) and can handle being pelted by micrometeors, yet it is thin enough to meet the necessary lightweight requirements. The material was tested at the Jet Propulsion Laboratory in Pasadena, California, and at Wright-Patterson Air Force Base in Dayton, Ohio.

The sails will be sent into space in a small container, then deployed when their location is reached. The side of the sail that faces the sun is covered with highly reflective material. The sail is thus transformed into a gigantic mirror, 200 meters by 200 meters for outer planet flight, 1,000 meters by 1,000 meters for interstellar flight. The force generated by the sun on the smaller sail is about the same as the weight of a first-class letter. However, this very small force is continuous, and there is no friction or resistance in outer space. Over hours, days, months, and years, this tiny amount of acceleration results in a speed to overtake space exploration vehicles like *Voyager* and *Pioneer* at the far reaches of the solar system.

Once the ship travels beyond the influence of the solar winds, it will continue coasting at an incredibly high speed beyond the solar system. The Planetary Society, which was founded by scientist Carl Sagan among others, hopes to send a test probe into space using solar wind technology. Their ship has 30-yard sails and costs about $4 million. If the first flight is successful, they hope to someday send a ship to the stars. Scientists have calculated that a ship propelled by solar sails would reach the nearest star in approximately forty years.

• • •

While solar sails aren't the same as a cosmic surfboard, it's quite conceivable that both work on the same principle. The Silver Surfer appears to be an outrageous comic book idea at the moment. However, a few hundred years from now when we have conquered space and teens are looking for a new thrill, the Surfer might not seem so impossible. Or even unusual.

13

The Case of the Missing Antimatter
Sinestro

When **DC Comics** editor Julius Schwartz revived the superhero known as Green Lantern during the Silver Age of Comics, he provided the character with a new origin and background. No longer was the Green Lantern a supernatural being powered by an ancient lamp from the Far East. Instead, Hal Jordan was an Earthman who came across a dying member of the intergalactic Green Lantern Corps and was recruited to take his place. Using a concept very similar to that of E. E. Smith's famous science fiction series, the *Lensmen*, the Green Lantern corps was a league of 3,200 heroes from across the universe. The Green Lanterns used power rings given to them by the Guardians of Oa to preserve peace and harmony in the galaxies.

Unfortunately, no matter how exacting the screening process, there's always the chance that one less-than-deserving individual will make it through all the tests and challenges of the Guardians and become a Green Lantern. With 3,200 Green Lanterns roaming the universe of billions of stars armed with near godlike powers, it's actually pretty amazing that more of the Corps aren't evil. Still, one is enough, since one very determined, self-centered, cruel, and ambitious Green Lantern nearly succeeded (a number of times!) in destroying the Corps and the Guardians. His name was Sinestro, and for many years he was the most dangerous enemy of Earth's Green Lantern, Hal Jordan.

Sinestro was a perfect example of Lord Acton's famous saying, "Power tends to corrupt and absolute power corrupts absolutely." As Green Lantern for the planet Korugar, Sinestro protected his people by isolating them from the galaxy. Any invading forces were mercilessly attacked and driven off. He forbade the people of Korugar from using space travel, preferring to keep the population under tight control in their home world. He ruled the planet as an absolute dictator, having dissolved the planet's high council, and anyone who even slightly disobeyed one of his commands was severely punished.

The Guardians of Oa trusted Sinestro and never thought to check on how he was performing his duties. Instead, as his sector of the galaxy appeared peaceful, they gave him the job of training a new Green Lantern, Hal Jordan of Earth. It was Hal, after a number of complicated adventures, who discovered that Sinestro was a power-crazed madman willing to do anything to maintain his idea of peace and order. With the help of a number of other Green Lanterns, Hal defeated Sinestro in a long series of battles. When brought before the Guardians of Oa, Sinestro was stripped of his power ring and banished to the antimatter universe of Qward. But evil never rests in comics.

Like most Green Lanterns, Sinestro was a very determined individual and almost impossible to kill. He contacted the rulers of the antimatter universe, the Weaponers of Qward, and made a deal to help them in their attempt to conquer the matter universe. The Weaponers gave Sinestro a yellow power ring and sent him back to the matter universe to stir up trouble. Again and again, the Guardians and Hal Jordan imprisoned the turncoat Green Lantern, but no matter how hard they tried, Sinestro always managed to escape. For major evildoers, the Weaponers were remarkably tolerant of Sinestro's continued failures, considering how many times he tried to destroy the Green Lantern corps and failed. If Sinestro had been a member of the Soprano Mafia family, he would have been whacked a long time ago. Sinestro was to the Green Lantern as Luthor was to Superman and the Joker was to Batman—a foe who refused to admit defeat and returned repeatedly over the years with yet another plan to conquer his nemesis. And like those other foes, Sinestro never succeeded.

While Sinestro is without question one of the most powerful villains in the DC universe, we dealt with his powers and their source of energy in *The Science of Superheroes* when we discussed Green Lanterns in general. Though a renegade and a traitor, Sinestro still commands pretty much the same powers as any Green Lantern and draws his energy from the giant cosmic battery of the Guardians. Sinestro is a mirror image of Hal Jordan, and while an interesting character, the science behind his existence is old news.

However, Sinestro's hideout in the antimatter universe of Qward is another topic altogether. While many cosmologists believe in antimatter, they're having a hard time finding it. And, since the universe isn't composed of matter and antimatter in equal amounts, scientists are having an even harder time explaining what happened to the missing antimatter. Which is a question even the Guardians of Oa might have a hard time answering.

Modern physicists have developed a quantum field theory that's consistent with both quantum mechanics and the theory of special relativity. They call this theory the standard model of particle physics. It deals with the fundamental particles that make up all matter in the universe, as well as the strong, weak, and electromagnetic forces that deal with them.[1]

The standard model separates all matter into six particles known as quarks and six particles known as leptons. These fundamental particles interact by exchanging force carrier particles such as photons. Quarks were first conceptualized by physicists Murray Gell-Mann and George Zwieg in 1964, when they theorized that all matter was made up of three fundamental particles. Gell-Mann named these elementary particles quarks, taking the name from James Joyce's novel *Finnegan's Wake*.[2]

In the years that followed, scientists confirmed the existence of quarks through experimentation and discovered that there were six different quarks. Each one became known as a flavor. The two lightest quarks were named *up* and *down*. The third quark flavor was named *strange*. Scientists gave it that name because of the strange long life of the K particle, the first particle discovered that contained the particular quark.

The fourth flavor was discovered in 1974 and was named *charm* for no particular reason. The fifth and sixth quarks were originally named *truth* and *beauty*, but scientists finally decided cute names were getting out of hand. The fifth and sixth quarks were then named *top* and *bottom*.

The other type of matter particles are the six types of leptons. Three leptons have an electrical charge and three do not. The three charged leptons are the electron, the muon, and the tau. The muon and the tau are charged like an electron but have much greater mass. The three other leptons are the three types of neutrinos. Neutrinos have no electric charge and very little mass.

Taken together, quarks and leptons make up all matter that we have discovered in the universe. Energy is transmitted between these particles by force carrier particles: the photon transmits electromagnetic force, the gluon transmits the strong atomic force, the W and Z particles transmit the weak atomic force, and an undiscovered gravity particle tentatively named the graviton transmits gravity.

How exactly is force transmitted between quarks? Imagine a boy and a girl standing on a slippery waxed floor of a gym. The boy is holding a ball and throws it with all his strength to the girl. She catches it, but needless to say, doing so causes her to slide back several feet. The boy has also moved back a foot in exerting the energy throwing the ball. Energy has been transmitted from the boy to the girl in the form of the ball, and they both have been affected by the energy transmission. Now just think of the boy and the girl as quarks, turn the ball invisible, and call it a photon.

The standard model was proposed in the 1970s. Experiments conducted throughout the 1980s verified the standard model's assumptions to incredible precision. Every particle predicted by this theory has been found. The standard model is considered one of the breakthroughs of modern physics, as it explains in just about every way how the universe works.

Unfortunately, the standard model has a few problems, the most serious one being that it doesn't deal with the fourth fundamental force, gravity. Unifying gravity with quantum mechanics is a major element of string theory. Cosmologists are working hard to include gravity in the standard model in what would then be a complete theory of fundamental interactions.

There's another even more annoying problem regarding the standard model. It deals with antimatter, or more importantly, the lack of it in our universe. As mentioned earlier, if there's no antimatter, then how would the antimatter universe of Qward exist? Where would Sinestro go when exiled from our universe?

To understand the scope of the problem, we first need to understand exactly what antimatter is and what it is not. In 1928, British scientist Paul Dirac combined the theory of special relativity and quantum theory into what became known as the Dirac equation. This equation provided a description of elementary particles possessing half-integer spin such as the electron. However, instead of one answer, the Dirac equation had two solutions. One answer described the electron, which has a negative charge; and the other answer described a particle similar to the electron but with a positive charge. At the time, no such particle was known to exist. However, the positive particle was discovered in cosmic ray showers in 1932 by physicist Carl Anderson and was labeled the positron, or the antimatter particle of the electron.

Antimatter and matter particles are exactly the same except for a few properties. The most important difference between the two is that their electrical charges are reversed. The antiproton, for example, is the antimatter equivalent of the proton. The two particles are identical other than the proton has a positive charge and the antiproton has a negative charge.

Being exact charged opposites, particles of matter and antimatter cancel each other out in a rather explosive manner. Combining one gram of matter and one of antimatter would be equivalent to detonating 10,000 tons of TNT. Which leads to a problem at the moment the universe began.

Turning back the clock 15 billion years brings us to the instant of the big bang. All energy and elementary particles in the universe are squeezed together into a sphere perhaps the size of a basketball. Needless to say, with the entire universe compressed into such a small amount of space, the inside of the sphere is incredibly hot.

The instant the big bang occurs, particles of matter and antimatter expand in an incredible rush of energy. As the universe starts to cool, around the time scientists estimate is 10^{-43} seconds after the explosion, equal amounts of matter and antimatter are formed. This is generally accepted as true, since one of the basic concepts of the standard model is that the universe is symmetrical. Which means in simplest terms that for every particle of matter created, a matching particle of antimatter is created. However, as already mentioned, equal amounts of matter and antimatter combine and explode.

So if equal amounts of matter and antimatter were created, they would have destroyed each other and created vast amounts of light and energy, leaving a vast amount of energy but *no* matter in the newly created universe. Or, in other words, the big bang would have resulted in a nonexistent universe.

We know that's not the case because we're here and made of matter. So something needs to be corrected in the standard model to account for the extra matter that formed the universe and resulted in us being alive billions of years later.

Now, the victory of matter over antimatter at the moment of the big bang can be explained easily by just stating that there was a minor difference in symmetry between matter and antimatter. Still, most cosmologists find that explanation lacking (i.e., cheating). If you're going to explain how the universe formed and can describe in great detail how everything worked after the first nanosecond but can't provide an explanation for what occurred before, then the description isn't complete. Nor is it very satisfying—that is, the explanation is inelegant.

Famous Russian nuclear scientist and Noble Peace Prize winner Andrei Sakharov proposed that the answer to the problem might be an elusive effect known as CP violation. CP violation states that matter and antimatter don't always behave in the same way, resulting in a one-in-a-billion imbalance in favor of ordinary matter. Even that difference would suffice to destroy all the antimatter in the universe and form enough matter to create all the galaxies and cosmic clouds in the universe.

The mathematical concept of symmetry is extremely important in particle physics. Right and left hands are symmetrical: mirror reflections of each other. Plus, you can change one to the other by flipping over either hand. Fundamental particles can be described in a similar way.

Matter and antimatter particles are symmetrical, as they have the opposite signs for several properties, most notably, electric charge. Particle theory describes this relationship by a "mirror" operator, C. In other words, working on a fundamental particle like an electron with C yields the antiparticle, a positron.

Another "mirror" operator, P, also applies to fundamental particles, this time in space instead of by charge. This property is comparable to flipping a right hand to the left hand. P changes the sign of a property called parity, which Paul Dirac proved is opposite for particles and antiparticles.

When particles interact, if the signs for C and P add up for all the particles the same as before they interacted, then C and P are each said to be conserved. But if the signs are different, we have CP violation. The first time CP violation was observed was in 1956 when scientists studying "strange" mesons called neutral kaons discovered that they sometimes decayed into two pions (another subatomic particle), a CP-violating situation.[3]

Ever since that time, scientists have studied the decay of subatomic particles in hopes of proving that CP violation is why matter exists in the universe. Scientists at the Stanford Linear Accelerator (SLAC) in Menlo Park, California, continually test elementary particles, hoping to find other inconsistencies in the standard model. These physicists hope that the subatomic particle, the B-meson, might help them in their quest.

B-mesons are heavy particles, and when such particles decay, they shed mass, turning into lighter ones. Physicists at SLAC create B-mesons and then track them in the accelerator's giant tunnels. They hope to discover some small difference between the path predicted for the B-meson by the standard model and its actual path. That hasn't happened yet, but scientists are still looking.

Taking symmetry a step further is a theory popular with many physicists today called leptogenesis. Instead of examining quarks at the instant of the big bang, leptogenesis deals with leptons, particularly neutrinos. In leptogenesis, heavy neutrinos turn into antineutrinos during the first nanosecond after the big bang. When this happens, matter is forced into existence to balance out the abundance of antineutrinos. Cosmologists consider leptogenesis important, as it is one of the few cosmological theories dealing with dark matter—the unseen matter that makes up much of our universe.

Meanwhile, other cosmologists and theoretical physicists are hard at work developing a new theory called supersymmetry. This concept suggests that a number of undiscovered subatomic particles exist. Along with each particle having an antimatter duplicate, supersymmetry predicts that there's yet another duplicate. These other massive particles are known as shadow particles. Thus, for every quark, there also exists a squark. For a gluon, there's a gluino, and for an electron, there's a selectron. Supersymmetry ties in with string theory, which relies on the belief that we live in a universe of ten dimensions.

Unfortunately for the Weaponers of Qward, none of the major cosmological theories being researched at present offer any hope of an antimatter universe. A thread common to the standard model, symmetry, CP violation, supersymmetry, and even leptogenesis is that all of them deal with why there is little or no antimatter in our universe.

Still, some cosmologists believe that the universe could be composed of regions of matter and antimatter. These cosmologists think there could be entire galaxies made up of antimatter. Most astronomers disagree. Outer space isn't empty. Antimatter galaxies would collide fairly frequently with intergalactic clouds of hydrogen gas made up of normal matter. These collisions would produce huge explosions, resulting in massive amounts of gamma rays, which would be easily detectable on Earth. That's never happened.

Other physicists argue that antimatter might exist and we just don't have the right tools to locate it. Any antimatter particles like antinuclei entering our atmosphere would be immediately destroyed, as they would combine with nuclei in the air and convert into energy. Thus, these scientists argue that the search for antimatter must be conducted above the atmosphere of Earth. In 1998, a high-energy particle detector was flown on the *Discovery* space shuttle for ten days. During that entire time, not one antinucleus was identified among the more than three million nuclei measured by the detector. Still, plans are in the works for an advanced version of this detector, an alpha-magnetic spectrometer, to be installed in the international space station. If any antimatter is entering our atmosphere, this device should detect it.

Is the antimatter universe of Qward out there or does it exist only in the minds of the writers who chronicle the adventures of the Green Lantern? What happened to the antimatter that was created at the instant of the big bang? Is leptogenesis or CP violation the key to understanding the standard model? These are questions that at least for now only the Guardians of Oa can answer.

14

Crisis on Infinite Earths

The Silver Age of Comics began in March 1956 with *Showcase Comics* #4 edited by Julius Schwartz. *Showcase* served as a testing ground at DC Comics for new heroes before they were awarded their own books. If the *Showcase* issues featuring a certain character sold well, then the superhero joined the DC lineup with his own book. If sales slumped, the character disappeared. The character featured in *Showcase* #4 was the Flash.[1]

The Flash was a Golden Age DC Comics hero whose original run had ended years earlier. Schwartz decided to bring him back, but this time with a much more scientific, believable origin. The new Flash, police scientist Barry Allen, proved to be very popular, and not only got his own comic book series but encouraged Schwartz to continue reviving past superheroes with new science fiction–style origins. Major characters included the Green Lantern, the Atom, Hawkman and Hawkgirl, Aquaman, and others. At the same time, new, entirely original characters started to appear as well, as the market for new comics swelled with the baby-boomer generation.

Things got somewhat more complicated in 1961 when Gardner Fox, a comic writer who had worked during both the Golden Age and Silver Age of DC Comics, wrote a story titled "Flash of Two Worlds" for *Flash Comics* #123. In this story, the Silver Age Flash, Barry Allen, meets the Golden Age Flash, Jay Garrick. Fox, a veteran comic book scriptwriter and sometimes contributor to the pulp magazines, had a simple but logical explanation for the existence of two Flashes. They lived on parallel worlds, each in their own separate universe different only in the rate of atomic vibration. The concept was accepted without complaint by fans who found the interaction between the Golden Age superhero and his modern counterpart vastly entertaining.

In comics, like all of publishing, success breeds imitation. In DC's case, the success of one Golden Age superhero soon led to another and another. The Justice League of Earth found itself working in partnership with their 1940s counterpart, the Justice Society of America, which existed, like the Golden Age Flash, on what became known as Earth 2. The popularity of the Justice Society revival started wheels turning at the DC office. If one parallel universe featuring alternate versions of DC characters was popular, why not feature another? Perhaps one inhabited with entirely different heroes and villains? Or comic book superheroes bought by DC from other companies when they went out of business?

Years passed and the number of parallel worlds featured in DC Comics kept growing. After a while, it became difficult to tell which character, or which manifestation of a character, came from the original Earth or was a doppelganger from another alternate Earth. It finally became apparent at DC Comics that the parallel worlds concept had lost its charm and that there were too many characters and their look-alikes in the comic book universe. Something had to be done to reduce the confusion.

The solution was *Crisis on Infinite Earths*, a twelve-part crossover miniseries created by Marv Wolfman and George Perez that appeared from January to December 1985. A crossover in comics is defined as a superhero from one comic guest starring in a comic featuring another superhero as the main character. In this series, every DC superhero crossed over sometime during the year to appear in the saga.

All of the characters from the DC parallel universes and some new characters were forced to join together in a gigantic battle against a villain called the Anti-Monitor. The Anti-Monitor (who naturally was opposed by the Monitor) schemed to destroy all the alternate worlds of the DC universe, which he dubbed the *Multiverse*—a term we believe was first used by Michael Moorcock in his Elric series—leaving only his universe, the Anti-Matter universe, intact. In some unexplained manner, the energy generated by the destruction of universes made the Anti-Matter universe more powerful. The Anti-Monitor ruled the Anti-Matter universe. Destroying the Multiverse left him with no competition as absolute dictator of the cosmos.

The purpose of *Crisis* was to cut down the number of superheroes in the DC universe, get rid of the numerous parallel Earths that had been created during the past several decades, and trim back some of the excessively complex character histories that had developed over the years. The ambitious series did exactly that, wiping out every parallel universe except the original Earth and killing off many minor and a few major characters including Supergirl and Barry Allen, the Silver Age Flash. The series also changed the history of a number of other heroes and villains. Needless to say, while many readers were pleased with the changes, many others were not. The editors and writers at DC spent several years fine-tuning the results of *Crisis*, including reviving several dead characters and rewriting the history of a number of others.

Still, *Crisis on Infinite Earths* was a huge success for DC. The series was the first major company-wide crossover that resulted in permanent changes to a comic book universe. Characters died, worlds perished, and universes were destroyed. Multiple versions of major superheroes ceased to exist. And the Multiverse changed back into a single, much more manageable universe.

There was only one problem, which was ignored by the series' writers and editors. The concept used to create and justify the Multiverse also made it impossible to destroy the Multiverse. *Crisis on Infinite Earths* was a great story, but it never happened.

As best as can be determined by science fiction historians, the first story featuring a parallel universe was the novel *The Heads of Cerberus* by Francis Stevens, which appeared in magazine form in 1919. In almost a century since then, parallel universe stories have become a staple of science fiction. The basic concept of these stories is that at critical points in history two parallel universes are created with two possible outcomes—for example, one in which the Nazis lost World War II and a second in which the Nazis won World War II.

Parallel universe stories usually don't focus primarily on the event that creates the new universe but the history that follows. Popular stories center on what the world would have been like if the South had won the Civil War or what the world would be like now if the American Revolution had failed and the United States was still a British colony. While not all parallel world stories involve warfare, the concept is very popular with military science fiction writers, since wars very often feature important moments when a decision one way or another will have a profound effect on the future. An interesting parallel world novel is *Bring the Jubilee* by Ward Moore, where the main character goes back in time to change history by making sure the South doesn't win the Civil War.

As the field grew more sophisticated, people realized that it's not as easy to pick out these branch points in history. Oftentimes, the most minor events can change the world just as easily as a major battle. For example, if Adolf Hitler had gotten sick during the great pneumonia outbreak in 1919 and died, the history of the twentieth century would have been much different. If the New York City snowstorm of 1888 had not occurred, paralyzing traffic for days in the city, the New York subway might never have been built. The importance of the most minor occurrences may not be noticeable for decades or even centuries.

In Ray Bradbury's classic story, "A Sound of Thunder" (adapted by EC Comics for their sci-fi line), a man who time-travels back to the age of dinosaurs changes modern history by accidentally stepping on a butterfly millions of years in the past. Charles Harness postulates a world without violence by altering the behavior of early cavemen in his novel *The Paradox Men*. H. Beam Piper imagines the church controlling the secret of gunpowder. With the incredible growth in alternate history stories in the past few decades, hundreds upon hundreds of stories and novels explore the many worlds of "what if." Which raises the question of exactly how many parallel universes exist.

Obviously, there are a lot. Even if we just consider the major events in history as moments when the universe branches into two new realities, the number is staggering. How many major events have occurred in human history? Thousands, tens of thousands, hundreds of thousands, millions? Plus, every time a new reality is created, events in its future will also result in parallel worlds. And this branching effect has been going on since the beginning of history. So the number of branches is in the billions of billions.

It's also important to note that every event in prehistory (before humans became the most prominent life-form on Earth) needs to be taken into account as well. As demonstrated by the Bradbury story, the smallest change in the Jurassic Age could have changed all of history afterward. The first volcanic eruption on a newly formed Earth might have had an important impact on the air we breathe now. Therefore, every event that ever happened on Earth needs to be taken into account when studying parallel worlds. The universe isn't selective or intelligent; nor does it decide which events are important to the future and which are not. The number of parallel universes that exist in direct relationship to our own world is a function of the number of events that have taken place since the creation of Earth.

Amazingly, this notion ties in closely with one of the major theories regarding quantum mechanics: the many worlds interpretation as proposed by Hugh Everett III. According to Everett's theory, whenever multiple possibilities exist in quantum events, the world (i.e., the universe) splits into many worlds, one for each possibility. As in science fiction, these worlds are all real and exist simultaneously with the first, while remaining unobservable by any of the others.

Everett's theory is not well liked by many physicists who prefer the Copenhagen interpretation of quantum mechanics developed by

Niels Bohr and Werner Heisenberg in the late 1920s. The Copenhagen interpretation says that measurement outcomes are basically indeterministic in quantum mechanics. Albert Einstein was a strong opponent of the Copenhagen interpretation, expressing his doubts in the famous line, "God does not play dice."[2]

While the Copenhagen interpretation of quantum mechanics is fascinating, we'll stick with the more interesting many worlds theory. Working with that concept, we find that the number of alternate universes for Earth, although incredibly large, is finite. Since Earth has not existed forever, if we had a gigantic computer and a lot of spare time for calculating, we could come up with the number of all possible quantum events that have taken place since Earth was formed. We could thus calculate every possible parallel universe created by those events. From there, we could track down every possible quantum event that took place in all these branch universes. Continuing outward, following every possible branch, counting quantum events at the speed of light, we could record billions upon billions of probability worlds that are linked to the first quantum action on planet Earth. The number would be mind-boggling.

Still, the sum of an immense but finite group of numbers is a finite number. So, though the voyage would be staggering, since the number of universes created by Earth over its billions of years of history is finite, we could travel from the first created parallel universe to the last. Science fiction stories that discuss the details of traveling across millions upon millions of parallel universes include work by Poul Anderson, Andre Norton, H. Beam Piper, and Keith Laumer.

But is that immensely huge number the total of all parallel universes that exist? By now, anyone reading these books knows that we never ask a question without having an unpleasant answer ready. In this case, we've dealt with the many worlds theory as it relates only to Earth. However, Earth is just one planet, part of one solar system, part of one galaxy, part of one galaxy cluster, part of our universe. Since the entire universe contains atoms whose particles are subject to the laws of quantum mechanics, every atom in the universe is subject to the many world theory. While the life cycle of planets, stars, and even galaxies is finite, our current theory about the universe states that it began with a big bang billions of years ago and has been expanding ever since, creating new stars, new solar systems, and new galaxies.

As we understand it, the universe is infinite in size. Thus, we have an infinite number of atomic particles whose movement creates parallel universes throughout the entire universe—that is, since there is an infinite number of atomic particles, there is an infinite number of parallel universes. After all, the title of the story is *Crisis on Infinite Earths*. Now it's time to learn exactly what infinity means. And to discover why the very mention of the word makes the Anti-Monitor break out in tears.

The word *infinity* has been part of our vocabulary for the past hundred years, but ask a hundred people to explain what it means and you'll get a hundred different answers, most likely all of them wrong. Everyone agrees that infinity has something to do with a number that's bigger than any other number, but that's about as far as they get. Things get really confusing when you ask, "What do you get when you cut infinity in half?"

At the end of the nineteenth century, Russian-born mathematician Georg Cantor founded the set theory branch of mathematics, which made working with infinity possible. Using the concepts of set theory, Cantor was able to describe infinity and discover many of its properties. Set theory served as the foundation of many of the important mathematical discoveries of the twentieth century. Though derided for his ideas during his lifetime, Cantor was recognized after his death as one of history's most brilliant mathematicians.

Cantor described a set as a collection of well-defined and well-distinguished items considered as a whole. Thus, a group of postage stamps is a set, as is a collection of baseballs or a group of numbers. A collection of unrelated objects is a set {man, woman, phone book, 7}, since the items in the set are all well defined and can be distinguished from each other. (For later use, we will call this set A.) The numbers from 1 to 1,000,000 form a set for the same reason. Sets can be large or small. Most important, sets can be finite or infinite. A finite set has a finite number of objects. No matter how many objects are in a finite set, given enough time, you could count all of the items. An infinite set is one where no matter how long you spend counting the objects, you will never finish.

Working entirely with sets, Cantor needed a new method to prove that one set was bigger than another or that two sets were the same size. Cantor named the operation he used to compare the sizes of sets a one-to-one correspondence. In simplest terms, a one-to-one correspondence pairs up the objects in one set with the objects of another set. If we take set A, we can pair each object in the set with the numbers {1, 2, 3, 4}. In doing so, one object of each set is paired with one object of the other set, no object is left out, and no object is paired up with more than one item. By setting up this direct correspondence, we are sure that both sets {man, woman, phone book, 7} and {1, 2, 3, 4} have the same number of objects.

Setting up a one-to-one correspondence describes through set theory how we count things in a finite set (such as how many tomatoes are in a bag or how many people live in New York City). However, using a one-to-one correspondence also works when comparing infinite sets. Which is where things get interesting.

The set of all whole numbers, also known as the integers {1, 2, 3, 4, 5, 6, . . .}, is an infinite set known as aleph-null. It has no end because no matter how big a number we select as the biggest, we can always find a number bigger by just adding one. And so on forever.

Using the concept of one-to-one correspondence and the basic definition of infinity, let's answer the question we posed earlier: What's bigger—infinity or one-half of infinity? Instead of trying to answer this question using the basic principles of arithmetic, we'll find our answer using Cantor's methods. If we take aleph-null (the infinity of the integers) as our beginning set, then half of aleph-null would be found by canceling out half the numbers in the set. Or, in simpler terms, if aleph-null is the set $\{1, 2, 3, 4, 5, 6, 7, 8, \ldots\}$, then one-half of aleph-null is $2, 4, 6, 8, 10, 12, \ldots$. To discover which set is bigger than the other, we arrange the two sets in a one-to-one correspondence.

1	2	3	4	5	6	7	8	$\ldots n \ldots$ (where n is any number)
2	4	6	8	10	12	14	16	$2n$

We've constructed a one-to-one correspondence between the set of all the whole numbers and the set of the even numbers. Each set is infinite, but every number from the first set (n) has one and only one corresponding number in the second set ($2n$). Which means that the infinite set of the integers and the set of even numbers are the same size, making the set of even numbers an infinite set as well.

We can go through the same proof using odd numbers to show that the set of odd numbers is the same size as the set of the integers. What about the set of all numbers that are multiples of the number 5?

1	2	3	4	5	6	7	8	9	10	$\ldots k \ldots$
5	10	15	20	25	30	35	40	45	50	$\ldots 5k \ldots$

Because we can set up a one-to-one correspondence between the set of integers and the set of numbers that are multiples of five, they are both the same size and are both infinite.

What if you add two infinite sets together, forming a new set? Would that set be bigger than either of the two sets? The answer hopefully is fairly obvious and is why we picked the infinity of odd and even numbers as sets, since combining those two infinite sets merely results in the infinite set of all counting numbers. Adding an infinite set to another infinite set of the same size will still result in the same-size infinite set, as will subtracting an infinite set from an infinite set. Multiplying an infinite set by 2 will result in the same-size infinite set.

The same holds true when multiplying the infinite set of counting numbers by any finite number. Or, even more astonishing, multiplying the set times itself. Infinity times infinity results in the same-size infinity. Cantor defined infinity in the following manner: a collection is infinite if some of its parts are as big as the whole.

At first, Cantor believed that there was only one size infinite set. Then he wondered if perhaps the set of rational numbers might be bigger than the set of the integers. However, he came up with an ingenious proof to show that the integers and rational numbers could be put in a one-to-one correspondence.

To do this, let's examine the rational numbers. A rational number is defined as a number that can be expressed as p/q, where p and q are both integers and q is not zero (since division by zero is not allowed in our set). Thus, rationals can all be expressed by fractions made up of integers. If we like, we can write rationals as $1/3$, $2/5$, $7/3$, $4/2$, and on and on. Many rationals can be expressed as several different but equal fractions; for example, $2/4 = 3/6 = 7/14$, and so on. That doesn't really matter. What does matter is that we can assemble the infinite set of rational numbers in a very logical order so that they can be put into a one-to-one correspondence with the integers, meaning the integers and the rationals are the same-size infinity.

Constructing our set is easy. We draw a number grid and use our rational numbers to locate the points on the grid that identify each line and column. (And so on and so on, both across and down.) This field will produce all of the rational numbers except zero. By using our grid pattern, we have demonstrated that every rational number can be paired up with an integer. Our proof is complete, since adding a zero to an infinite set does not change the size of the set.

After proving that the rational numbers were of the same-size infinity as the integers, Cantor set out to investigate the real numbers—those numbers that could not be represented by a fraction represented by p/q, where q is not zero. Cantor determined that these other numbers, known as the real numbers, formed an infinite set bigger than the integers. His proof is known as Cantor's diagonaliza-

1/1   1/2 —— 1/3   1/4 —— 1/5   1/6 —— 1/7   1/8 . . .

2/1   2/2   2/3   2/4   2/5   2/6   2/7   2/8 . . .

3/1   3/2   3/3   3/4   3/5   3/6   3/7   3/8 . . .

4/1   4/2   4/3   4/4   4/5   4/6   4/7   4/8 . . .

5/1   5/2   5/3   5/4   5/5   5/6   5/7   5/8 . . .

6/1   6/2   6/3   6/4   6/5   6/6   6/7   6/8 . . .

7/1   7/2   7/3   7/4   7/5   7/6   7/7   7/8 . . .

8/1   8/2   8/3   8/4   8/5   8/6   8/7   8/8 . . .

tion proof, and it is one of the greatest mathematical proofs ever done. It is famous not only because of its amazing conclusion but also because of its astonishing simplicity.

Consider all of the numbers between zero and one. Consider only those points that cannot be expressed by a rational number. How big is that set of numbers? It is an infinite set of numbers, since it never comes to an end. To see if it is the same order of infinity as aleph-null, we need to show that the set of these (real) numbers can be put in a one-to-one correspondence with the integers—that is, the whole numbers.

Cantor suggested we take the list of all real numbers between 0 and 1. Such a set would look something like this:

.	1	2	3	4	5	6	7	8	9	...
.	2	7	1	8	2	8	1	8	3	...
.	7	9	2	6	1	4	6	3	8	...
.	3	1	4	1	5	9	2	6	5	...
.	4	4	2	9	7	2	5	1	8	...
.	1	2	0	2	4	5	6	1	4	...
.	9	5	1	3	8	2	7	7	3	...
.	4	1	4	2	1	3	6	9	9	...
.	1	2	1	6	2	3	2	4	3	...

We note that these real numbers (or irrational numbers as they are called) continue to infinity and that our list is infinitely long. As laid out, this set could be put into a one-to-one correspondence with the integers. Every irrational number could match up with one, and only one, integer.

Now, let's look at our group of numbers again, but this time highlight the diagonal of the set:

.	**1**	2	3	4	5	6	7	8	9	...
.	2	**7**	1	8	2	8	1	8	3	...
.	7	9	**2**	6	1	4	6	3	8	...
.	3	1	4	**1**	5	9	2	6	5	...
.	4	4	2	9	**7**	2	5	1	8	...
.	1	2	0	2	4	**5**	6	1	4	...
.	9	5	1	3	8	2	**7**	7	3	...
.	4	1	4	2	1	3	6	**9**	9	...
.	1	2	1	6	2	3	2	4	**3**	...

We've just created a new infinity irrational number: .172175793 ... This new number we will call X. X is infinitely long and is between 0 and 1, so it fits into our set. Now, let's create a new number, Y, based on X. If one of the digits of X is a 1, we will make it a 2. If any digit of X is a number other than 1, we will make it a 1. We do our substitution and come up with our number Y being .2112111111 ...

Now, examine *Y*. It is different from any number on our list in one place (the diagonal) if not more. Therefore, it does not belong to our set, since it is a different number than any number listed. Thus, it is not matched in a one-to-one correspondence with any number in that set. Which demonstrates that when we try to establish a one-to-one correspondence between the integers and the real numbers, the real numbers are in a bigger infinity than aleph-null. The infinity of the real number is known as aleph-one. Thus, not only does infinity exist but different sizes of infinity exist. Though we only need aleph-null to bedevil the poor Anti-Monitor.

Return with us to the DC Multiverse, the scene of *Crisis on Infinite Earths*, just a few minutes before the superpowerful Anti-Monitor sets his evil plan in motion to destroy all the normal matter universes and leave only the Anti-Matter universe. Which, of course, he plans to rule for all eternity. Or so he/she/it (it's never made very clear in the story just exactly what the Anti-Monitor is other than being very big, very powerful, and very ugly) thinks. How sad to realize the Anti-Monitor is doomed to defeat. Not by the combined efforts of all the superheroes in the Multiverse—in the comic book series describing their valiant struggle against the Anti-Monitor and his invading shadows, the greatest superheroes of all the Multiverse were barely able to preserve one Earth, one universe, from this arch-fiend. However, in the real everyday world, these superheroes will never be put to the test. The Anti-Monitor faces a much more frightening enemy—mathematics.

Everything in the Multiverse takes time. The movement of an electron around the nucleus of an atom takes time, even if it is measured in billionths of a second. Destruction, especially thorough destruction, takes a fairly good amount of time. Judging from the examples shown in *Crisis*, it often took the Anti-Monitor and his minions days to defeat the numerous superheroes who united to save their own particular universe from destruction. Days to wipe out a universe of infinite size? Even that statement raises some questions about what sort of weapons were being used.

Consider a universe the size of our own. It's estimated using our best telescopes that it consists of a billion galaxies, each containing a billion or more stars. The Anti-Monitor comes from the Anti-Matter universe, and his weapon of choice is shooting antimatter particles at matter. When particles of matter and antimatter collide, they are both destroyed and turned into energy. If the Anti-Monitor came to our universe, he first would need an infinite number of antimatter shells to fire at the infinite number of stars in our universe. Also, antimatter is matter (as opposed to energy), and there's no evidence that antimatter travels faster than matter. Therefore, antimatter particles are bound by the same laws of physics as matter and cannot travel faster than the speed of light. So antimatter missiles fired at star systems a hundred billion light-years away are going to take a hundred billion years to reach their destination. That doesn't exactly correlate with the Anti-Monitor's claim that he's been able to destroy nearly the entire Multiverse in a fairly short amount of time.

Maybe, just maybe (for the sake of argument), the Anti-Monitor has an arsenal of gigantic mother-of-all-universe bombs that he uses to destroy parallel universes. One bomb explodes per universe and somehow breaks down all the laws of physics in that universe, reducing everything into energy in a minute or so. Sounds impossible, but at least it would be more practical than firing antimatter missiles at everything. That would surely cut down on the invasion time of the Multiverse. Or would it?

Let's see. Remember, we're dealing with infinity, and infinity can cause major problems. The Anti-Monitor wants to destroy all the parallel universes in creation, and it takes one minute to destroy each one. So our equation is easily defined as Time = (1 minute/universe) $\times$ (∞ universes) = ∞ minutes.

Before we take a close look at our answer, let's realize that the time to destroy a universe has no effect on our equation unless it is zero. Zero times anything is zero, but destroying a universe in zero time isn't possible. It has to take some amount of time, even if that time is measured only in seconds, milliseconds, or nanoseconds. And, hopefully by now it is obvious from our brief study of infinity that whatever time length we plug into the equation, the result will be the same. Because any positive number multiplied by infinity results in infinity as the answer.

So it would take the Anti-Monitor an infinite amount of time to destroy the Multiverse. Though it's clear from the story that his shadow allies cannot destroy a universe entirely on their own, let's assume for a second that they could. So we divide our infinite time by 1,000 helpers. Or divide the time by 1,000,000 helpers. Does it change anything? Not in the least, because infinity divided by any whole number (merely reversing infinity multiplied by any whole number) is still ∞.

Let's stick with our original premise that the Anti-Monitor can destroy a parallel universe in a minute, but that because there is an infinite number of parallel universes, it is going to take the Anti-Monitor infinite time.

Exactly how long is ∞ time? Well, we know if we divide ∞ by 60 minutes/hour, the result is ∞ hours. If we then divide ∞ by 24 hours/day, and then by 365 days/year, our result is still as expected. Time = ∞ years. Which in a moment of extreme hubris, the authors express in a formal statement we call Weinberg's corollary: infinite work cannot be completed in finite time.[3]

Of course, the Anti-Monitor could start the job, but he would never get it done. After all, while he is busy destroying alternate universes one minute at a time, new parallel worlds are continually popping into existence due to the movement of electrons in the infinite universes that remain. Even if the Anti-Monitor began his evil scheme fifteen billion years ago, slightly after the big bang, he'd still be spending all his time destroying things. Fifteen billion years is a very long time. But it's still less than infinite years. Remember, infinity means never having to say you're finished.

Crisis on Infinite Earths was a great idea for straightening out the continuity of DC Comics. But someone should have checked a little further on what the title really implied. *Crisis on A Lot of Earths* would have worked, but Crisis on Infinite Earths, never.

15

Frustration in Five Dimensions
Mr. Mxyzptlk and Bat-Mite

Superman and Batman have faced hundreds of foes in their course of more than sixty years of fighting evil. They've battled master criminals, superpowered aliens, sentient machines, lunatic inventors, beautiful seductresses, and just about every possible evildoer imaginable under the sun. They've defeated them all, from Lex Luthor to the Joker to Bizarro to Catwoman to Mr. Freeze. They've succeeded in stopping the worst menaces devised by the most diabolical fiends in the universe. Almost. For despite all their successes, all their triumphs, there are two unbeatable villains who at best Superman and Batman have fought to a draw. A pair of imps whose very names give pause to the greatest heroes of the DC universe: Mr. Mxyzptlk and Bat-Mite.

While it has never been positively established that these beings both come from the same world—the fifth dimension of Earth—it seems fairly likely that they inhabit the same sector of the cosmos. Both Mr. Mxyzptlk and Bat-Mite share powers that seem magical to normal humans. They can bring inanimate objects to life, travel from place to place in the blink of an eye, become invisible, float on air, and seemingly create something out of nothing. Yet the two mischievous imps always proclaim that they are using science, not magic, to perform their tricks. Such claims are probably best encapsulated by science fiction master Arthur C. Clarke in his third law: "Any sufficiently advanced technology is indistinguishable from magic."[1]

The technology used by Mr. Mxyzptlk and Bat-Mite is not only more advanced than anything used on Earth but is five-dimensional in construction. Thus, by its very definition, this technology is impossible to define logically on these three-dimensional pages in our three-dimensional terms. Nor could we ever understand how such machines worked, being mentally bound by the three-dimensional space in which we live. Fortunately, though we can't see, sense, or visualize the fifth dimension, we can learn a lot about it from basic mathematics. And hopefully we can learn some of the secrets of its inhabitants as well.

Mr. Mxyzptlk appears for the first time in *Superman Comics* #30 (September 1944) in the lead story titled "The Mysterious Mr. Mxyzptlk."[2] The adventure begins with a small, odd-looking man reading a newspaper, not watching where he is going, walking into the middle of a busy street. Seconds later, he's hit by a truck. Ambulance workers look for a pulse but can't find one. After putting the body on a stretcher, they try to lift it into the ambulance but can't. It's too heavy. Others try to help, to no avail, when the strange man suddenly sits up and states, "Confusing, aren't I?" He flies into the ambulance, drives it up the side of a building, then into the atmosphere, where it explodes.

The next sequence of the story has the same little man looking for his friend, McGirk. A statue comes to life and leaves with the little man. Afterward, the man appears at a swimming pool, seemingly drowning in the deep end. But whenever lifeguards try to save him, he appears somewhere else, even floating above the water. After a few more oddball episodes in which water splashes out of radios and music broadcasts from refrigerators, Superman finally confronts the strange character on the top of a bridge.

Mxyzptlk explains that he comes from another dimension. As a scholar and court jester, he discovered two special words: one that would transport him to our world, the other that would send him back. Mxyzptlk is amazed that Earth's science is so backward, and, of course, he plans to conquer the world.

"So what's the second magic word?" asks Superman.[3] Mxyzptlk is so amused that Superman would ever think he would reveal the magic word that he says it aloud, "Kltpzyxm"—his own name reversed—and is sent back to his own dimension.

The second Mxyzptlk story appeared in *Action Comics* #80 (January 1945) a few months after the first. Titled "Mr. Mxyzptlk Returns," the adventure is pretty much a rewrite of the first adventure, with the fifth-dimensional imp causing chaos in Metropolis and Superman reversing the disasters. As in the first story, Superman uses a pretty simple trick to send Mxyzptlk back to his home dimension.

The Mxyzptlk stories proved an amusing break from Superman's relentless battles with master criminals and evil scientists and appeared throughout the 1940s and 1950s. When Julius Schwartz gained control of the Batman line in the late 1950s, one of the first things he did was look around for successful gimmicks used in other DC superhero stories. Thus, it wasn't long before Batman and Robin met their own fifth-dimensional pest. In *Detective Comics* #267 (May 1959), the lead story announced "Batman Meets Bat-Mite!"

The story begins with Batman and Robin going to the Batcave at the beginning of a midnight patrol. They notice that some of their equipment has been disturbed but don't have to look far for the intruder. He's a tiny, elflike creature dressed in a shrunken version of Batman's outfit. The imp announces, "I come from another dimension where all men are my size! Batman, I've observed and admired your exploits for years, so I decided to help you fight crime with my unearthly powers. Won't that be FUN!"[4] The other-dimensional imp has already given himself a name: Bat-Mite.

Bat-Mite, needless to say, proves to be more of a headache than a help for Batman. Like Mr. Mxyzptlk, Bat-Mite possesses the ability to bring inanimate objects to life. After a long, tiring week, Batman finally convinces Bat-Mite to return to his own dimension. The imp leaves but promises to return sometime so that they can have more "fun" together.

Bat-Mite kept his promise and returned a number of times in the 1960s to pester Batman and Robin all in the name of having a good time. Like Mr. Mxyzptlk, he served as comic relief from the gallery of oddball criminals who Batman fought most months. Bat-Mite was amusing if taken in small doses.

Based on the theory that two disasters are better than one, Bat-Mite and Mr. Mxyzptlk teamed up several times: in "Bat-Mite Meets Mr. Mxyzptlk" in *World's Finest Comics* #133 (November 1960) and in "The Incredible Team of Bat-Mite and Mr. Mxyzptlk" in *World's Finest Comics* #143 (February 1962). When the history of the DC universe was totally revised in *Crisis on Infinite Earths* in 1985, both Mr. Mxyzptlk and Bat-Mite survived. Since they are fifth-dimensional beings, the destruction of parallel universes didn't alter their world in the least. In the past few years, the two imps have continued their crazy battles with Batman and Superman. In perhaps their most memorable encounter, "Last Imp Standing," Mr. Mxyzptlk and Bat-Mite manage to destroy for a second time every parallel universe and superhero encountered in *Crisis*.[5]

Explaining technology so advanced that it is mistaken for magic is beyond the skill of this book's authors. However, much of Mr. Mxyzptlk's and Bat-Mite's power is a direct result of their coming from the fifth dimension. As ordinary humans, we are aware of only three physical dimensions: height, width, and length. Duration (or time) is classified by Einstein as the fourth dimension. What is the fifth dimension, and are Mr. Mxyzptlk and Bat-Mite possible? The answers might surprise you.

The easiest way to define multidimensional space is by using basic geometry.

A point (.) has no length, width, or height. It is just a point, defining a singularity in any dimension.

A straight line segment

_____

defines one dimension. The line segment has length but no width or height. It can be measured by the distance between a starting point and an end point (which is why it is a segment). A straight line segment can be measured (e.g., it is 2 inches long), whereas a straight line continues to infinity.

Two line segments at right angles to each other and attached at a single point (the vertex or corner) define a two-dimensional space. The space defined by these two lines is known as a plane. Line segments in a plane are called sides, or edges. They are one-dimensional but are in a two-dimensional space. The end points of edges are known as corners, or vertices. They have no dimensions. In two dimensions, every vertex serves as the end point of at least two edges. If the edges are segments, every vertex is the end point of two sides, and none of the sides cross; they form a polygon, which is a multisided object that has both length and width. Polygons have the same number of sides as they have vertices in two-dimensional space.

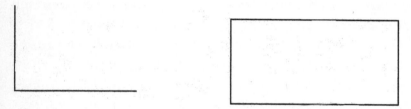

Two lines at right angles form a plane.   A four-sided polygon.

A polygon often divides a plane into two sections: an inside and an outside. The inside is called the region enclosed by the polygon, and the area of this region is called the area of the polygon. When two line segments meet at a vertex, they form two angles: one inside the figure, called the interior angle, and the other outside the vertex. A circle, of course, is a two-dimensional figure without any angles or vertices.

A circle in a two-dimensional plane.

If two line segments at right angles define two dimensions, then it seems logical that three line segments, each at right angles to each other (think of the corner of a room), define three dimensions. A three-dimensional figure has width, length, and height. It is often called a solid. The surface regions of a solid are two-dimensional planes, while the end points of the edges are still called the vertices and are zero-dimensional.

If all the faces of a solid are planes, every edge has two faces, every vertex is the vertex of at least three faces, and no two faces cross each other (actually simpler than it sounds), the figure is called a polyhedron. As with polygons, polyhedra divide space into two pieces: an inside and an outside. The inside of a polyhedron is called the region enclosed by the figure and is measured in volume. Everything in our world of three dimensions is a three-dimensional object whose height, width, and length can be measured. Any point in three-dimensional space can be defined as a group of three numbers. The point $(3, 5, 7)$ would be defined as three spaces over on the x axis, five spaces up on the y axis, and seven spaces out on the z axis.

If we ignore Einstein's mention of duration as a dimension, then defining a four-dimensional space merely requires putting four line segments at right angles to each other. Points in this dimension will be defined by four spacial coordinates. A five-dimensional space, such as the one Mr. Mxyzptlk comes from, would be a higher-dimensional world, which is defined by five line segments all at right angles to each other and by five spacial coordinates. Such spaces are called Euclidean spaces. In mathematics there are non-Euclidean spaces, but fortunately they are not important to our discussion of the fifth dimension.

Thus, the fifth and the fourth dimensions exist, at least in mathematical terms. But where are they? How can we be sure that they're not just some imaginary concepts put forward by mathematicians to keep the public puzzled and math majors employed? Aren't there any examples of four-dimensional space, and if there are, why can't we see them?

In simplest terms, the fourth dimension is defined by taking four straight line segments and placing them at right angles to each other. Please feel free to try this at home in your leisure time. It's impossible in our universe. Why? Because our minds and our senses are three-dimensional, and while we can discuss a fourth dimension in abstract mathematical terms, we cannot imagine a four-dimensional world. We cannot see, hear, touch, or in any other way encounter that fourth straight line at right angles to the other three. As three-dimensional creatures, we are totally blind to the fourth dimension. Does this mean that the fourth dimension doesn't exist?

Definitely not. One of the results of Einstein's general theory of relativity was to prove that space was curved in four dimensions. That proof would take an entire book to explain (and probably would leave us all in the dark), but the basic idea behind it can be easily explained by an example.

It's a two-dimensional world, with just length and width. Still, it's a comfortable world for the creatures who live on it: a bunch of circles we will call the O family. The O's can only move in two dimensions; they can't raise themselves off the paper, but to them, up is not a dimension.

The O's at home.

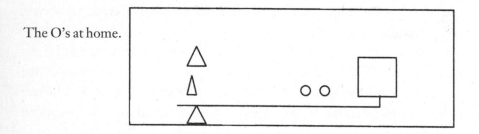

One day, the O's decide to go on a walk. They decide to walk to the forest at the end of their farm. Since the shortest distance between one point and another is a straight line, they travel in a straight line from their home to the edge of their land. So they take the path to the forest, traveling in a straight line in a two-dimensional world.

However, the O's don't realize that their two-dimensional world is actually curved in three dimensions. As they always remain flat on the surface of their world, they can't detect the curvature in space, but it exists. In many ways, the O's are quite similar to three-dimensional people walking on Earth. While we know Earth is round, taking a short walk on a street doesn't convince us that the world is curved.

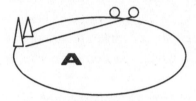

World seen from a sideways (three-dimensional) view.

Now the O's are walking to the forest in what to them appears to be a straight line. However, it is not a straight line but a geodesic—the shortest line between two points that is confined to the surface. Obviously, the shortest line between the O's and the forest is the straight line **A,** which runs through the ovoid. But the O's are creatures of length and width only. They cannot sense, much less travel in, a third direction: height. Thus, they don't realize that their world is curved or that there is a shortcut through a higher dimension that would take them to the forest more quickly.

Keeping the O family in mind, move this example from their two-dimensional world to our three-dimensional one. We're in the starship *Surprise,* and we want to travel from our sun to the Alpha Centauri sun. We know that these two suns are more than four light-years apart (billions upon billions of miles). At a speed close to the speed of light, the trip will take us more than four years if we follow a straight-line course between the two suns. However, we believe in general relativity and know that it proves that space is curved in four dimensions, and we remember the O's. We turn on our "warp" drive (not yet invented), and it burns a hole into the fourth dimension. We fly along the fourth-dimensional shortcut to Alpha Centauri, and it only takes a few weeks. Surprise!

Most characters in science fiction novels, movies, and television shows use warp drives to travel. In reality, these drives are merely engines that somehow break down the undefined barrier to the fourth dimension. That's a lot different than flying around at warp 8, but it doesn't violate any of the laws of physics (which sci-fi shows routinely mangle). Unfortunately, none of these shows explain how characters ascend into the fourth dimension or navigate when they are in it; nor do they descend out of it. It's magic as defined earlier in this chapter by Arthur C. Clarke.

Assume now that Mr. Mxyzptlk comes from a fourth-dimensional world. He can walk entirely in his own dimension, never actually entering the other three dimensions below his, but he can see and hear everything that is taking place there. He can pop in at any location he wants, seemingly from nowhere, then pop out, back into nowhere, which is really just his fourth-dimensional perch.

As a fourth- or fifth-dimensional being, his body most likely extends into the higher dimension but is not visible to dwellers on three-dimensional worlds. Only the three-dimensional parts of his body are visible in our world. Thus, pretending to die is a lot easier for the imp, as it's impossible to give him a complete and thorough examination.

Magic is science if most of what takes place is unseen by the audience. Bringing statues and signs to life is a lot easier if the props extend into the unimaginable fourth dimension and are attached to machines that make them mobile. Remember, the fourth dimension is not invisible in the usual sense. It's impossible to detect because our three-dimensional senses can only see three-dimensional items. We only see parts of four-dimensional buildings, machines, and people. Which is one of the main reasons that the tricks played by Mr. Mxyzptlk and Bat-Mite seem so astonishing. Like any good magic trick, we're unable to see what's hidden behind the curtain.

Could Mr. Mxyzptlk and Bat-Mite actually exist? Forget the funny names and outfits for a minute and revise the question. Can beings from the fourth or fifth dimension exist and visit our world? After we've denied the existence of so many superheroes and supervillains, can we honestly say that oddball creatures with the ability to pop up out of nowhere, wield astonishing powers, then vanish without a trace really exist? To which we can answer only, maybe.

The truth about the matter is that we know as little about higher dimensions as we do about the farthest galaxies in the universe. Please don't make the mistake of confusing dimensions with parallel universes (as discussed in chapter 14 on *Crisis on Infinite Earths*). These higher dimensions are not alternate worlds created by some major military event or some small chance happening in quantum mechanics. They are the place where space is curved and people can move from point A to point B without traveling the distance between. Unfortunately, knowing what they are like and showing they exist are problems.

On June 10, 1854, Victor Riemann gave a lecture in Germany entitled "On the Hypotheses Which Lie at the Foundation of Geometry," in which he used mathematics to describe possible "higher, unseen dimensions." Riemann suggested that the universe was four-dimensional and that the basic laws of physics were much simpler in that fourth dimension than in our three-dimensional world.

Riemann's lecture caused a major stir among scientists but was soon forgotten by everyone but psychics and philosophers, who seized on the concept of a higher dimension to validate all sorts of unconventional beliefs. Unfortunately, many people wanted physical proof of the fourth dimension, and Riemann couldn't offer it. If this dimension existed, no one had gone there and returned. Riemann's equations were relegated to the world of pure mathematics until 1919, when a little-known mathematician, Theodr Kaluza, proposed that the fourth dimension actually existed but was "smaller than an atom."[6]

In 1926, another mathematician, Oskar Klein, looked at Kaluza's ideas using the newly invented theory of quantum mechanics. Klein came up with the theory that this new dimension described by Kaluza was "Planck length," the smallest size allowed in quantum mechanics: 10^{-33} centimeters. Unfortunately, since no machine existed at the time that could work with objects this small, the Kaluza–Klein theory remained just a theory.

But physics has an odd way of surprising scientists. In 1968, there was a sudden rush to examine hyperspace again with the development of the superstring theory. According to this theory, fundamental particles are viewed as hyperspace vibrations of incredibly small multi-dimensional strings. Suddenly, Kaluza–Klein was hot again. The superstring theory unified all the known forces in the universe in a specific number of dimensions, quite often defined as ten.

There's only one problem: all ten dimensions are contained in the same unreachable and unexaminable Planck length. The superstring theory, which may be our best explanation yet for how the universe works, can't be tested.

You can almost hear Mr. Mxyzptlk laughing—

Comic Writers Tell All

1. Please tell us about your connection to the comic book field.

RICHARD CLARK, artist: Like almost everyone working in comics, I got into the making of them because I read them voraciously as a kid. My absolute favorite was the *Incredible Hulk* (though I ended up getting hooked by the *Uncanny X-Men* and the *Fantastic Four* a little later). Thus, it was a terrific boon to me to work with Mark Texiera on the *2000 Incredible Hulk Annual* as one of my first credited jobs in comics. I got to do quite a lot of penciling and inking on that book and had a terrific time working with and learning from Tex.

From there I went toward more independent publishing, getting some great opportunities to work with the good folks at Moonstone Books. With them I drew two books: *Jack Hagee*, written by C. J. Henderson, and *Vampire the Masquerade: Giovanni*, written by Robert Weinberg.

I've also hatched the mad idea that I'd self-publish my own character, Mün Hed. So far, I've distributed nearly two thousand free ashcans, put up a very rudimentary site (www.munhed.com), and am busting my butt to bring the first issue to print. All this, of course, is taking place while I maintain a robust illustration business that services clients like HBO, *Playboy*, the *Wall Street Journal*, and the New York Yankees, to name just a few.

JEFF MARIOTTE, writer and editor: I've worked in almost every part of the comic book industry—in retail almost thirty years ago, much more recently as VP of marketing at WildStorm Productions, then as senior editor when WildStorm was acquired by industry leader DC Comics, then as editor-in-chief at IDW Publishing. I've also written a lot of comics, most notably *Desperadoes*, which was nominated for a Stoker Award and an International Horror Guild Award. Recent work has included some comics based on the TV shows *CSI* and *The Shield* and a series about vampires working for the CIA called *CVO: Covert Vampiric Operations*, for which the scientific explanation is effectively, *Hey, they're vampires, okay?*

JENNIFER CONTINO, editor and critic: I'm a lifelong comic book fan and enthusiast. Several years ago, I began hosting some of the "Official Chats" that DC Comics had online and started writing for Wizard's online site, SequentialTart.com, and Fandom. From there, I landed "offline" work in *Wizard, Comic Book Marketplace, Comicology, Comics Spotlight, Newtype USA, Publisher's Weekly*, and several other print magazines. In July 2002, I decided to quit my day job working with abused children and work full-time as a writer. The following month I began working for *The Pulse*, www.comicon.com/pulse, a daily comics and anime/manga news site. I was chosen as an Eisner judge for the 2003 Eisner ceremony.

KURT BUSIEK, writer: I've been a professional comics writer since 1982, working on characters as varied as Vampirella, Mickey Mouse, and Jell-O Man. I'm probably best known for my work on *Marvels*, a look at the Marvel universe through the eyes of an ordinary human, and *Astro City*, a series about what it's like to live in a world of superheroes and deal with the extraordinary intruding into ordinary life, both of which have won quite a few industry awards, including multiple wins for best writer, best series, best single issue, and more. These days, in addition to continuing *Astro City*, I'm writing *Conan, Arrowsmith* (a fantasy series about a World War I infused with magic), and other projects, and have recently completed *JLA/Avengers*, a long-awaited crossover series featuring the top heroes from both Marvel and DC Comics. Outside of writing, I have never been a professional rodeo rider but was one hell of a paperboy, and have also been a short-order cook, a literary agent, an editor, a sales manager, and more.

ANDREW PEPOY, writer and artist: I've always had an interest in comics of some sort, from watching reruns of *Batman* and *Looney Toones* cartoons after school and doodles of Superman with crayons, to collecting *Peanuts* books and picking up a few *Wonder Woman* comics during the days of the Lynda Carter TV show, to my full-on collecting starting when the first Superman movie came out when in fourth grade. Soon after that, I started drawing the characters in the comics and started creating my own. I was fortunate enough to have a nearby library with lots of comics history books and reprints of old comic strips like *Flash Gordon*, *Buck Rogers*, *Terry and the Pirates*, and more. So, from early on I was just hooked on comics of all kinds and doomed to a life creating them. I began publishing my own fanzines when twelve, sold my first pro work when twenty, and was working for Marvel and DC by twenty-one while still in college. In the years since then, I've inked hundreds of comic books for many major publishers, working on such characters as Superman, Spider-Man, the X-Men, Godzilla, GI Joe, Scooby Doo, the Simpsons, and many more. I've also penciled a few comics (*FemForce*, *Heroes Anonymous*, etc.), as well as redesigning and drawing the *Little Orphan Annie* newspaper strip for a year. I also created/write/draw my own independent creation, *The Adventures of Simone & Ajax*, the story of a twentyish girl and her small dinosaur, which has appeared in various anthologies and one issue of it's own title.

2. Do you think real science has a place in today's comics?

RICHARD: Science has been studied since its earliest pursuits in an effort to explain the real, physical workings of the world in which we live. As such, real science has a place in any type of popular entertainment, including comics, because people relate better to stories taking place in a believable world. The wealth of information about how the real physical world works available at a glance to even the casual reader of comics dictates that the creators of comics have to know enough about it to fib convincingly at the minimum. If not, the tales we spin run the risk of alienating readers, as the readers would be able to poke holes too big in the stories to suspend their disbelief.

JEFF: Fortunately for comics, almost anything goes these days—although to get beyond straight superhero comics you might have to look somewhat beyond the mainstream, best-selling comics. I'm sure there is science in comics, even if you discount Jim Ottaviani's comics that are specifically about science and scientists.

What's questionable is how much of an influence real science has on mainstream comics. I suspect there's a lot more fake science—in the early days of Image Comics, there was a definite prejudice against having heroes and villains with magic-based powers, so there was a premium put on coming up with convincing-sounding "scientific" explanations for the bizarre powers of the heroes and villains. I was guilty of that myself, with a limited series I wrote called *Hazard*, about a bounty hunter who was infected with a nanotech virus, which was controlled by his worst enemy—if he didn't do the bad guy's bidding, the nanotech would replicate itself to the point that it overran his system and killed him, but as long as he "behaved," it gave him incredible powers.

I haven't seen much recently that makes me think that's changed—science tends to be used as an explanation for what is essentially magical, or inexplicable. Of course, real science is complicated enough that it might be difficult to explain in the course of a thirty-two-page story and still have room for the rest of the story.

JEN: I know that when I was in third or fourth grade I had a high school reading level and knew about several subjects, because of comics. I definitely believe science has a place in comics and would rather have a scientific explanation for powers than the easy-out "magic" answer. Atom, Batman, Reed Richards—they used to constantly explain in analytical terms or scientific terms the reasons for specific things happening. I miss that. I wish we could have it back. Outside of superhero comics, I think the success of projects like *Dignifying Science* and *Fallout* from Jim Ottavianni, as well as *Clan Apis* from Jay Hosler, prove that people are interested in reading stories that have a definite scientific slant.

KURT: Sure. I think comics is a storytelling medium like any other, and as such there's a place in it for any kind of story. Some of those have only a passing acquaintance with the laws of physics, while some, like Jim Ottavianni's nonfiction comics (including *Two-Fisted Science*, *Dignifying Science*, and *Fallout*), take their science very seriously. One of my favorite comics of recent memory is *Clan Apis* by Dr. Jay Hosler, a comic about the life cycle of a honeybee.

ANDREW: While I haven't seen much real science in comics lately, I'm sure, when needed, it has a place. Not long ago I was contacted by a well-known (at least in science circles) physicist who thought comics would be a great tool to teach kids about science and wanted to develop a whole line of science teaching and history comic books. So comics could be used in the service of science in a very direct way. Until this project actually develops, though, while comic books of today seem to have all sorts of technology and scientific miracles in them, I doubt much of it is based in real science. The terminology is derived from real science, but from there it's all made up. After all, comics are a form of escapism for many readers, and if there's too much real science (and other real facts) in the comics, they lose their purpose to that audience.

3. Do you think there is more or less science in comics now than there was during the Silver Age of Comics (the late 1950s and 1960s)?

RICHARD: The types of advances being made in science during the Silver Age of Comics—deepening awareness of atomic sciences, landing a man on the moon, etc.—made science-based stories much more prevalent than they are today. While science continues to explore the physical universe, the discoveries and advances are much less spectacular than they were forty years ago, so the mass appeal isn't as prevalent. I mean, how much cooler is it to send people into space for the first time than it is to meditate on quantum physics? And we must always remember—comics, like rock 'n' roll, makes its bread and butter in all things cool.

JEFF: I don't know that the science in the Silver Age was real science either—there seemed to have been a strong emphasis on science then—the '70s saw a return to magic, fantasy, and sword and sorcery themes, but before that, there were lots of futuristic, gleaming cities, and noble scientists like Jor-el, Superman's dad, as well as scientist villains like Lex Luthor, Dr. Sivana, and Otto Octavius.

I think there are fewer noble characters in comics overall these days, so fewer noble scientists. There may have been a bit of a dumbing-down, as well, resulting in fewer scientist villains. Villains today are often more thuggish, or if not, they are financiers or politicians—rich men and women we all love to hate.

JEN: I'm not sure. Being thirty-one, I've just read a few hundred SA comics, so I can't judge all of the SA by the dozens of superhero comics I've read. Like I said, I can remember Atom and Batman using scientific reasoning, and, of course Mr. Fantastic using scientific terms when I was reading comics in the late '70s/early '80s, but I'm not sure how that has lessened or increased in recent years.

KURT: It depends on the book, I'd say. A lot of Silver Age comics concerned themselves with science but played it out in ridiculous ways. These days, superhero fans particularly are more interested in the physics of superpowers, so those comics that do address science are more likely to be grounded in some sort of reality, but more writers probably choose to avoid scientific details rather than get them wrong. Still, there are plenty of comics throwing around buzzwords like nanotech, viral machines, and so on, just as radioactivity and exotic elements were omnipresent in the Silver Age.

ANDREW: It depends on whether you're talking about real or imaginary science. I think most so-called science in today's comics is mostly the creation of the writers, while in comics of the '60s there seemed to be more real science. Sure, there was all sorts of fake science in the *Fantastic Four*, *Iron Man*, and others, but I seem to remember that very often, especially in older DC titles, there would occasionally be little explanations as to how some miracle of science worked. I seemed to learn a good bit from it, as my parents, including my chemist father, seemed surprised at what bits of knowledge I picked up from comic books.

4. When writing, illustrating, or criticizing a superhero and supervillain comic book, do you ever do any research on the science involved in the story? Do you feel that the needs of the story outweigh the need of having believable science?

RICHARD: On a personal level, I never approach a story, either literally or pictorially, with the research first. However, I never shirk my duties in that regard if I feel there's an important story element that needs research in order to fib convincingly. As such, I'm always a story guy, first and last, but I do believe that verisimilitude is key to the telling of any story. That in mind, I don't see any separation of believability and story—they're joined at the hip. However, I don't think any story has ever been served in history by the overtelling of any factual material. Balance between fiction and the facts that support it must always be maintained.

JEFF: When writing the *Hazard* miniseries, I did a little perfunctory research on nanotechnology, but just enough to throw some language around, since it was quickly apparent that it would be many, many years, if ever, before the kinds of applications I was describing would be possible. Writing *CSI: Thicker Than Blood* and *CSI Miami: Smoking Gun*, though, I recognized that part of the appeal of the TV shows is that the science is "real" (albeit exaggerated in terms of the speed with which it can be done), so I did the research necessary to make sure the crimes were solved with real, available science. My general feeling, which I suspect is the prevailing view, is that the science needs only be believable to the real layperson and not to readers who are terribly scientifically literate, with the exception of cases such as *CSI* where real science is expected and an intrinsic part of the property.

JEN: I don't usually do a lot of research when writing a review of a comic book. I believe most comics are fantasy, so if there are things that can't be explained or that are explained in a different way than what would happen in real life, then I chalk it up to it's a comic. However, if Jim Ottavianni's *Dignifying Science*, which is supposed to be biographic and factual, had costumed superheroes in it or Marie Curie flying, I'd call him on that. But as far as superhero/fantasy/fiction goes, I don't criticize or research—that much—the science behind a story.

KURT: To be honest, I'm more likely to research history than science—I've always been interested in exploring the characters more than the powers, so I'll generally base my portrayal of superpowers on what's already been established about that character, whether it's scientifically defensible or not. For instance, I did a story a few years back that involved a group of ants receiving an urgent SOS from the Avengers and passing it on to Ant-Man—this is utterly ridiculous science, of course, since it involves ants both understanding English and transmitting the message, radio-like, from ant to ant until it reached Ant-Man. However, we've seen it happen in Marvel Comics before, so my take on it is that ants apparently do understand English in the Marvel universe. Ridiculous? Sure, but it's a place where the ridiculous can be true, from ancient astronauts being mistaken for gods to lycanthropy to ants that understand human languages. I find that charming, so I enjoy making use of it.

If I have real science in one of my stories, I try to at least fact-check it, so if, for example, Quicksilver races off to a destination that requires him to break the sound barrier, the story will incorporate a sonic boom. But I find that the stories I write don't often require such detail—not because I'm consciously avoiding it, but because the ideas I have don't lean that way. And this is, ultimately, a medium where a 180-pound man can hit another 180-pound man with an impact equal to several thousand tons of TNT going off, and neither man will even be lifted into the air; nor will the ground beneath them be affected, presuming they're braced and transfer the impact to whatever they're braced against. So if we're going to ignore physics as basic as Newton, getting more complex science "right" seems less like accuracy and more like window dressing.

ANDREW: When in high school and college, I took a good bit of science (biology, chemistry, physics, and pseudoscience alchemy), and I know my way around a library, so I could do the research needed to get my facts right, but I choose not to. In my own writing of *Simone & Ajax*, I'm usually trying to tell some fun and silly story. My characters, one of whom is a dinosaur—so I'm throwing scientific fact right out the window from the start—have adventures in many different times and places, but I never try to give any kind of explanation of how they got there. Let the reader either just enjoy the story or make up their own idea of how Simone and Ajax got to ancient Atlantis or the moon.

When a producer wrote a screen treatment for *Simone & Ajax*, he tried to explain how they got from their home to wherever they were going. Despite any other changes to my characters he made, that was the only one I strongly objected to. It ruined some of the sense of fun I wanted to leave to the reader's imagination. It hurt the concept to explain things scientifically.

In the *Simone & Ajax* story I wrote most recently, they go to the moon in a used, old rocket they buy with green stamps and run into trouble with the moon people. Sure, I know that there's a lot more to going to the moon and that you can't buy a personal rocket, and that there aren't people on the moon, but I choose to ignore those facts so as to write a fun story. Now, if I were illustrating someone else's story and needed to get some effect or piece of scientific equipment right, I'd go and do the research, but in the end I feel the escapism of the story is more important than any fact, scientific or otherwise.

5. Supervillains often seem to be very much involved in science. Many of them, in fact, are scientists or inventors or people involved in scientific research etc. Do you think there is a bias in comic books that portrays scientists or people who study science as geeks, often people without a social life, who are consumed with jealousy toward people who seem to have no interest in anything intellectual? Are supervillains a reflection of a cultural bias against high intelligence?

RICHARD: Popular culture looks up to so many folks of high intelligence and so much is made about the virtues of greater thought that I'd be hard-pressed to think there's a cultural bias against high intelligence. Albert Einstein is still revered by pop culture, Stephen Hawking is a pop culture icon, and we rail against politicians and political policies that are categorically stupid. Why, then, do so many villains seem to be cut from the Brainiac cloth?

I can't say for sure, but I have a guess. So much literary inspiration comes ultimately from the Bible in Western culture, and while I'm not a big Bible kind of guy, I do remember that it was the fruit of the tree of knowledge that was forbidden to Adam and Eve. Perhaps we're still laboring under the idea in the puritanically based culture that bad things come from too much knowledge. . . .

JEFF: It's possible, I suppose, that comic creators and fans want to believe in a class of people geekier than themselves, but that would be stretching the bounds of believability. I think the predominance of scientist supervillains can better be explained by the fact that superheroes are such powerful characters that for a villain to be perceived as any kind of threat, he or she needs some gimmick that could make them dangerous to the hero. Science can provide a villain with that edge, and rather than create villains who hire scientists to do their bidding, creators can simplify the process by making the villain and the scientist one and the same.

JEN: I don't think there's a bias any more than I think that all comic fans live in a basement or with their parents. Just as many heroes as villains were scientists: Peter Parker (Spider-Man), Reed Richards (Mr. Fantastic), Hank Pym (Ant-Man/Giant-Man/Wasp), Barry Allen (Flash), Ray Palmer (Atom), and several others who are geniuses who are not evil. I think the amount of scientists used as villains in the comics was a reflection of writers and creators thinking someone would have to be very smart and keen to do the types of things these villains did. With the Comics Code Authority, I think a lot of the other professions creators may have wanted to use were prohibited. But I don't think there's a bias.

KURT: I don't think so—after all, many superheroes are geniuses, too, from Reed Richards to Iron Man to Batman and others. Were the Silver Age Flash to exist today, he'd be taking his cues from *CSI*, as a forensic scientist. The X-Men work with fabulous technology created by their leader, Professor X. And even a lout like Hawkeye is engineer enough to design his own sophisticated gimmick arrowheads.

I tend to think that many supercharacters (both heroes and villains) are rooted in science, whether real or fanciful, because it provides an answer to the question: So where do the powers come from? Technology provides origins.

ANDREW: No, I wouldn't say there's a bias. While there are many mad scientists, such as Dr. Doom or Lex Luthor, as supervillains, there are also many heroes such as Mr. Fantastic of the Fantastic Four, Hank Pym as Ant-Man/Giant-Man/Yellowjacket, Ray Palmer as the Atom, and to some extent Batman with his scientific gizmo-laden utility belt or Green Arrow and his trick arrows. Science is used by both sides of good and evil to try to win the day.

6. Recent comics seem to be relying more on magic and fantasy to power supervillains than advanced science. Do you think there is any specific reason for that, or is it just that writers and artists are looking for easier ways to explain superpowers?

RICHARD: Quite a lot of culture observers and critics talk about the pendulum effect, where ideas at the opposite end of the pendulum's arc get tapped as one side's ideas get overexploited. If I had to wager a guess about the prevalence of powers based in mystic realms or arts, I'd say it stemmed from too many science-based ideas having been told in a row. I'd like to chalk it up to that old saw for another reason, as well—I'd hate to think that storytellers were getting lazy en masse. That'd be one sorry state of affairs.

In addition, and this segues into the question below, I'd theorize that in a culture where everyone has become more and more dependent on technological devices—everything from digital cable to cell phones to personal computers—that readers and creators of pop culture media get overinundated with technology. Maybe even more than the answers above, this new leaning toward magic-based solutions is simply a reaction to how commonplace high technology has become.

JEFF: I think there's a cyclical variation—in the late '60s/early '70s, magic-based comics were all the rage. Current trends in genre fiction are toward magic as well—fantasy far outsells most science fiction, and there are many more fantasy novels than hard science fiction novels published today. Popular movies include the *Lord of the Rings* trilogy and *Harry Potter*, but there is very little hard science in the movie theaters. Comics reflect that or are part of the same general wave.

JEN: I think there are only so many nuclear reactions, gamma bombs, radioactive spiders, and the like that someone can come up with. Magic is easy. You don't have to have too many explanations other than "magic."

KURT: I don't know that it's all that new—supercharacters have been mystically rooted going back decades, as well. It may simply be that comics creators are looking to vary the explanations, so while the Fantastic Four meet a new villain who's a sentient mathematical expression (in Mark Waid's first months on the book), the JLA face ancient god-villains. I don't know how much of that is a new trend—after all, back in the '70s, new villains included the JLA's science-based Construct (a disembodied intelligence created out of the complex web of electronic transmissions around the globe) and the Fantastic Four's magic-based Salem's Seven (a group of mystical menaces created by a city of warlocks).

ANDREW: In the past we lived in a society less confronted with the wonders of science on a constant basis, so those "miracles," compared with everyday life, seemed to be pretty fantastic and thus were part of the escapism of comics. However, nowadays, as we use personal computers, cell phones, and other gadgets that were once part of the fantasy of comics, some other kind of escapism is needed from our technology-heavy life.

As for myself, never seeming to be able to get my computer to do what I want, annoyed by cell phones, and angered by voice mail that loses messages, my choice for escapism is a story that doesn't include all the science that seems to complicate my life. Of course, the paradox of that is that I'll go watch movies on my VCR or DVD player and sometimes read comics on a CR-ROM or microfiche. As for looking for an easier way to explain superpowers, I think your average reader has picked up enough modern scientific fact that the gamma rays of the Hulk or white dwarf star of the Atom seem less believable today.

7. Our world is becoming more and more technology oriented. For example, you're all reading these questions via e-mail instead of on paper. Teenagers routinely play video games and study school subjects on the Internet. Do you see science becoming more or less important in comics? Do you think science in

comics will have to become more accurate because teenagers now know so much more science than they did before?

RICHARD: Getting back to some of the thoughts above, any science to which current stories refer has to have at the minimum a ring of truth, since contemporary readers have unprecedented access to real science. For at least the short term, I think there's going to continue to be a leaning away from science as a springboard for stories due to its prevalence in contemporary life.

Additionally, and this thought just occurred to me, the deeper the awareness of the physical universe, the denser the material becomes to digest. Not too many folks want to slog through all the implications of finer and more subtle scientific thought when they're coming to be entertained. Michael Crichton does a pretty good job of lightly illuminating the scientific details serving as underpinnings to his tales, but even he gets too dense for some pop fiction readers. And that's in novels. We've got to keep in mind that there's still a strong leaning in comics toward short, episodic periodicals running less than fifty pages, and there simply isn't enough time to fully articulate anything too dramatically dense while we're trying to cram as much story into our limited space.

JEFF: Teenagers *use* much more science than they did before, but it's probably speculative to think they *know* more. In fact, science education in high schools is probably less comprehensive than in times past. When science requirements can be met by taking Microsoft Word classes instead of chemistry or physics, it's likely that today's young people understand less about the scientific underpinnings of the world we live in, even though they are quite capable of using its tools. I don't believe that young readers today are any more demanding of realistic science than those of times gone by, and in fact I suspect it's just the opposite.

JEN: I'm not sure if it will increase or decrease. I think we'll always see within a team a smart, scientific type who can answer those strange or incredible questions. I'm not so sure we'd ever see a comic just about scientists, but you never know. And, although there are a lot of techie kids and teens, I still think the majority of them are just reading comics for escapism and aren't going to fact-check too heavily.

KURT: I honestly don't know. That teenagers are using more advanced technology doesn't mean they understand it any better than teens in the 1960s knew how televisions worked. I think that it's generally more important to get the cultural details right—a plot is more likely to involve the use of cell phones or the Internet because the readers know it's possible but won't explain how that technology works. The use of a PDA in a comic is no more science than the use of a city bus; they're both technology, but unless the story concerns itself with how and why they work, then they're simply cultural elements. My guess is that the window dressing will change—we're less likely to have our heroes encounter civilizations on Mars now that we know more about it—but that superheroes will, at heart, remain more fantasy creatures, invoking physics when it's convenient, than creatures of consistent, if fictional, science.

ANDREW: I don't think science will become any more or less important in comics. Some creators will choose to use science, some won't . . . some will stick to the facts, others will expand on them, most will ignore them. As for the teenagers, simply having a computer or cell phone doesn't mean they have any knowledge of what makes it work. I actually think today's teenagers have less knowledge of science than they did fifteen to twenty years ago when I was a teenager. Perhaps the simpler technology of a transistor radio or telephone made the science behind it easier to understand for someone average like me, while today's day-to-day items such as computers, DVD players, and microwave ovens are so far beyond the average person that they don't even begin to want to know what makes it go. With that in mind, I think you could actually get away with just as much stretching of the facts of science as you could before. The terminology has changed, but the imaginations mutating the real science and the readers who are willing to suspend their disbelief haven't, at the core, changed all that much.

Thanks again to our talented group of comic professionals for participating in our Q&A session and sharing their insights with us on science in today's comic books!

Notes

1. The Original Dr. Evil: Lex Luthor

1. Upon realizing that Lex Luthor wants to destroy him, Superboy thinks, "It's unfortunate Luthor's father, a travelling salesman, is rarely home. His son needs a father's guidance." It's clear to us that, being a homicidal maniac, Lex Luthor needs a lot more than his father's guidance. Would it really make a difference if Luthor's father got a job at the local hardware store?

2. Jack Williams, *The Weather Book*, originally produced and copyrighted by *USA Today* (reprinted by Random House, New York, 1997, 19).

3. The American Solar Energy Society, 2002, as reported on their Web site at www.ases.org/.

4. www.research.ibm.com.

5. www.pbs.org/wgbh/amex/bomb/sfeature/mapablast.html.

6. www.pbs.org/wgbh/amex/bomb/sfeature/mapablast.html.

7. Steven L. Hoenig, *Handbook of Chemical Warfare and Terrorism* (Westport, Conn.: Greenwood Press, 2002), 20.

8. Paul Overeiner, "Time Travel: It May Be Possible, But Don't Buy a Ticket Yet," *Jackson Citizen Patriot*, Michigan, Apr. 1, 1992.

9. Michio Kaku, "A User's Guide to Time Travel," *Wired*, Aug. 2003, 104.

10. As reported at www.pr.caltech.edu/media/Press_Releases/PR11646 .html.

2. The Villain in the Iron Mask: Dr. Doom

1. *Fantastic Four* #1, Nov. 1961, 11.

2. *Fantastic Four* #5, July 1962 (and later issues).

3. www.news.bbc.co.uk/1/hi/sci/tech/1112411.stm.

4. www.rnw.nl/science/html/robots000807.html.

5. Ibid.

6. www.sciencenews.org/20010630/bob8.asp.

7. Ibid.

8. www.dailycal.org/article.asp?id=7517.

3. Computer Supervillain or Village Idiot? Brainiac

1. Oddly, it is Lex Luthor (see chapter 1) who removes the brain tumor from the body of Milton Moses Fine, aka Brainiac.

2. We're taking the liberty of quoting from our earlier book, *The Computers of Star Trek* (New York: Basic Books, 1999).

3. Rodney Brooks, "Elephants Don't Play Chess," *Robotics and Autonomous Systems* (North Holland: Elsevier Science Publishers, 1990). See also Rodney Brooks, "New Approaches to Robotics," *Science*, Sept. 3, 1991.

4. A digital signal has two discrete voltage levels. An analog signal varies continuously between minimum and maximum voltages.

5. K. Eric Drexler, *Engines of Creation: The Coming Era of Nanotechnology* (New York: Anchor Books/Doubleday, 1986), 78.

6. See the article about the Next Twenty Years discussion in San Francisco, at www.wired.com/news/technology/0,1282,37117,00.html.

4. Feathers and Fury: The Vulture

1. *Spider-Man* #2, Mar. 1963, 2.

2. "How Airplanes Work," www.travel.howstufffworks.com/airplane.htm.

5. The Kiss of Death: Poison Ivy

1. W. P. Armstrong and W. L. Epstein, "Poison Oak: More than Just Scratching the Surface," *Herbalgram* vol. 34 (1995): 36–42.

2. W. L. Epstein, H. Baer, C. R. Dawson, and R. G. Khurana, "Poison Oak Hyposensitization: Evaluation of Purified Urushiol," *Archives of Dermatology* (1974): 356–360.

3. www.poisonivy.aesir.com/view/fastfacts.html.

4. Isadora B. Stehlin, "Outsmarting Poison Ivy and Its Cousins," www .fda.gov/fdac/features/796_ivy.html.

5. www.zanfel.com, a Web site devoted to the anti–poison ivy product called Zanfel.

6. Isadora B. Stehlin, "Outsmarting Poison Ivy and Its Cousins."

7. www.bio.umass.edu/immunology/poisoniv.htm.

8. www.aad.org/pamphlets/PoisonIvy.html.

9. *The Columbia Encyclopedia*, 6th ed. (New York: Columbia University Press, 2002).

10. Howard Hughes Medical Institute, "Pheromones Control Gender Recognition in Mice," Jan. 31, 2002, as reported at www.hhmi.org/news/dulac.html; citing original findings by Catherine Dulac of Harvard University, as extracted from *Science Express* and *Science*.

11. Howard Hughes Medical Institute, "Pheromone Receptors Need 'Escorts,'" March 7, 2003, as reported at www.hhmi.org/news/dulac2.html; citing original findings by Catherine Dulac of Harvard University, as extracted from the March 7, 2003, issue of *Cell*.

12. World Health Organization, *Chloroform Health and Safety Guide* (Geneva: World Health Organization for the International Programme on Chemical Safety, 1994).

13. Frederick B. Rudolph and Larry V. McIntire (eds.). *Biotechnology: Science, Engineering, and Ethical Challenges for the 21st Century* (Washington, D.C.: Joseph Henry Press, 1996), 12.

14. Lois H. Gresh, *TechnoLife 2020: A Day in the World of Tomorrow* (Toronto, Canada: ECW Press, 2001).

15. Michael Gross, *Travels to the Nanoworld: Miniature Machinery in Nature and Technology* (New York: Perseus Publishing, 1999), 177.

6. Groping for Power: Doctor Octopus

1. www.marvel.com.

2. www.oandp.com/news/.

3. Gregg Nighswonger, "New Polymers and Nanotubes Add Muscle to Prosthetic Limbs," *Medical Device & Diagnostic Industry*, Aug. 1999, as reported at www.devicelink.com/mddi/archive/99/08/004.html.

4. This copolymer material is used in blood storage bags.

5. Gregg Nighswonger, "New Polymers and Nanotubes Add Muscle to Prosthetic Limbs."

6. Lois H. Gresh, *TechnoLife 2020: A Day in the World of Tomorrow* (Toronto, Canada: ECW Press, 2001), 185–195.

7. See the Biomimetic Products, Inc., Web site at www.biomimetic.com/faq.html.

8. "Smart Materials," *Scientific American*, May 1996, as reproduced at www.sciam.com/explorations/050596explorations.html.

9. See www.endo.sandia.gov/9234/smas.html.

10. Michael G. Zey, *The Future Factor: The Five Forces Transforming Our Lives and Shaping Human Destiny* (New York: McGraw-Hill, 2000), 84. Zey is the executive director of the Expansionary Institute, a consultant to Fortune 500 companies and government agencies.

11. Chris O'Malley, "The Binary Man: Step One," *Popular Science*, March 1999, 64.

12. "Tech 2010," *New York Times Magazine*, June 11, 2000.

13. See www.sciam.com/askexpert/chemistry/chemistry6.html.

14. K. Eric Drexler, as quoted in Michael Gross, *Travels to the Nanoworld: Miniature Machinery in Nature and Technology* (New York: Perseus Publishing, 1999), 199. Gross has a doctorate in physical biochemistry.

15. See www.sciam.com/askexpert/chemistry/chemistry6.html.

16. "Tech 2010," *New York Times Magazine*, June 11, 2000.

17. Mark Anderson, "Mega Steps Toward the Nanochip," *Wired*, Apr. 27, 2001, as reproduced at www.wirednews.com/news/print/0,1294,43324,00.html.

18. Michael Gross, *Travels to the Nanoworld: Miniature Machinery in Nature and Technology*, 153. Carbon nanotubes have been the subject of many fascinating articles in scientific magazines and journals during the past few years. Gross provides a brief and excellent introduction to the subject in his book. He writes, "Historically, carbon nanotubes are a byproduct of the phenomenon that became known as the *fullerene fever*. After Wolfgang Kratschmer's group at the Max Planck Institute for Nuclear Physics in Heidelberg had described a recipe suitable for the mass production of these soccer ball–shaped molecules, fullerene chemistry spread epidemically (quite literally, as the increase in publications and citations was successfully modeled using the mathematical descriptions developed for epidemics). One of the first scientists infected was Sumio Iijima, who worked in the research laboratories of NEC in Tsukuba, Japan. While trying to optimize his procedures to produce fullerenes, he slightly changed the technical parameters in his discharge apparatus and was surprised to find that instead of soccer balls he obtained long and thin fibrils. Electron microscopy revealed these fibers to consist of concentrically stacked graphite cylinders, whose ends were capped with fullerene-like hemispheres. Like the fullerenes, these tubes represented a novel modification of carbon. As their diameters typically measured a few nanometers, they became generally known as *carbon nanotubes*, or less formally, as *buckytubes*."

19. Mark Anderson, "Becoming Your Own Hospital," *Wired*, Nov. 11, 2000, as reproduced at www.wirednews.com/news/print/0,1294,40120,00.html.

20. Mary Ann Swissler, "Microchips to Monitor Meds," *Wired*, Sept. 28, 2000, as reproduced at www.wirednews.com/news/print/0,1294,39070,00.html.

7. Leapin' Lizards: The Lizard

1. "Man Is Turning Himself into Lizard," *Ananova*, May 12, 2001. Erik's Web site is www.thelizardman.com.

2. Lawrence Altman, *Who Goes First?* (New York: Random House, May 1998), 100.

3. Anne LeBold, "What Stimulates Limb Regeneration?" *Research*, summer 1995.

4. Ibid.

5. Phil Sahm, "The Mystery of Regeneration," *Health Sciences Report*, summer 2003, University of Utah Health Sciences Center, 1.

6. Ibid., 2.

8. Clothes Make the Man: Venom

1. "Smart" clothes feature sensors and cameras, tdc/trade.com, Mar. 2000.

2. Larry O'Hanion, "'Smart' Clothes Sense Every Need," *Discovery News*, Dec. 18, 2002, 1.

3. Lori Valigra, "Fabricating the Future," csmonitor.com, Aug. 29, 2002, p. 2.

4. www.ananova.com/news/story/sm_827582.html?menu=news.scienceanddiscovery.inventions.

5. www.star.t.u-tokyo.ac.jp/projects/MEDIA/xv/oc.html.

6. Lori Valigra, "Fabricating the Future," 4.

7. Ibid., 5

8. Larry O'Hanion, "'Smart' Clothes Sense Every Need," 1.

9. www.icewes.net/projdetails.htm.

10. www.icewes.net/projdetails.htm.

9. Grodd Almighty: The Evil Super-Gorilla

1. William H. Calvin, "The Emergence of Intelligence," *Scientific American* (Oct. 1994): 100–107.

2. www.pbs.org.wnet/nature/koko.

3. Sally Boysen, Scientific American Frontiers, as reported at www.pbs.org/saf/1108/features/boysen.htm.

4. William H. Calvin, "The Emergence of Intelligence," 100–107.

5. Ibid.

6. T. W. Deacon, "Fallacies of Progression in Theories of Brain-Size Evolution," *International Journal of Primatology* vol. 11 (1990): 193–236.

7. www.geocities.com/CapeCanaveral/Lab/2948/evoerac.html.

8. www.en.wikipedia.org/wiki/Punctuated_equilibrium.

10. A Magnetic Personality: Magneto

1. In the comic, the message is in block letters, but Magneto's signature appears handwritten.

2. Retcon: www.wikipedia.org/w/wiki/phtml?title=Retcon.

3. www.encarta.msn.com/encnet/refpages/RefArticle.aspx?refid=761566543.

4. Particle Physics, Science-Park Info: www.science-park.info/particles/forces.html.

5. Lawrence Osborne, "Savant for a Day," *New York Times Magazine*, June 22, 2003.

6. www.hfml.kunnl/phystod.html.

11. Immortality for Some: Vandal Savage and Apocalypse

1. In a letter to Jean-Baptiste Leroy, 1789.

2. We know Cro-Magnons didn't live 50,000 years ago, but according to DC comics, they did!

3. William the Conqueror was born in 1028, and Genghis Khan wasn't born until 1167, but those facts didn't seem to bother Savage's biographers, so who are we to argue?

4. If this sounds complex, be thankful we are not covering the alternate future series, *The Age of Apocalypse!*

5. Ronald Bailey, "Forever Young," *Reason*, Aug. 2002, 4.

6. Ibid., 3.

7. Ibid., 8.

8. Ibid., 11.

9. "Immortality Becomes Real," *Pravda*, July 2, 2003.

12. Have Surfboard, Will Travel: The Silver Surfer

1. *Fantastic Four* #48, Mar. 1966, 7.

2. *Fantastic Four* #48, Mar. 1966, 8.

3. Actually, the Watcher disobeys his orders fairly often. In the Star Trek universe, he'd be cited for disobeying the prime directive.

4. *Fantastic Four* #48, Mar. 1966, 22.

5. *Fantastic Four* #50, May 1966, 10.

6. Deborah J. Jackson, www.solarsails.jpl.nasa.gov/introduction.

13. The Case of the Missing Antimatter: Sinestro

1. See chapter 10 for more information about the four basic forces of the universe.

2. James Joyce, "Three Quarks for Muster Mark!" *Finnegan's Wake* (New York: Penguin, 1999), 383.

3. "Strange" refers to the quarks making up the subatomic particle.

14. Crisis on Infinite Earths

1. See our entire chapter about the Flash in *The Science of Superheroes*.

2. www.nationmaster.com/encyclopedia/Copenhagen-interpretation.

3. Our apologies in advance to all mathematicians who have come to the same conclusion.

15. Frustration in Five Dimensions: Mr. Mxyzptlk and Bat-Mite

1. Arthur C. Clarke, *Profiles of the Future* (New York: Henry Holt), 1984.

2. In this story, his name was spelled Mxyztplk, but in all stories that followed, the Mxyzptlk spelling was used.

3. *Superman Comics*, #30, Sept./Oct. 1944.

4. *Detective Comics*, #267, May 1959.

5. Evan Dorkin and friends, *Superman & Batman: World's Funnest* (New York: DC Comics, 2000).

6. www.enterprisemission.com/hyperla.html.

Bibliography
and Reading List

1. The Original Dr. Evil: Lex Luthor

"The Einstein Connection," *Superman* #416, 1986.

"How Luthor Met Superboy," *Adventure Comics* #271, 1960.

"The Impossible Mission," *Superboy* #85, Dec. 1960.

"The Last Days of Ma and Pa Kent," *Superman* #161, May 1963.

"The Luthor Nobody Knows," *Superman* #292, 1975.

"Superman versus Luthor," *Superman* #4, 1940.

Braunstein, Samuel L., "A Fun Talk on Teleportation," Feb. 5, 1995, www .research.ibm.com.

Hoenig, Steven L., *Handbook of Chemical Warfare and Terrorism* (Westport, Conn.: Greenwood Press, 2002).

Kaku, Michio, "A User's Guide to Time Travel," *Wired*, Aug. 2003.

Mitchell, Edward Page, "The Clock That Went Backwards," *The Sun* (San Francisco), Sept. 18, 1881.

Overeiner, Paul, "Time Travel: It May Be Possible, But Don't Buy a Ticket Yet," *Jackson Citizen Patriot*, Apr. 1, 1992.

PBS, *The American Experience*, "Race for the Superbomb," www.pbs.org/ wgbh/amex/bomb/sfeature/mapablast.html.

Wells, H. G. *The Time Machine* (New York: Scholastic Books, 1978).

Williams, Jack, *The Weather Book* (New York: Random House, 1997).

www.ases.org. The home page for the American Solar Energy Organization, with detailed information on how to use solar energy for daily life.

2. The Villain in the Iron Mask: Dr. Doom

Fantastic Four Comics #1–100 (and other issues).

Bonsor, Kevin, "How Exoskeletons Will Work," *HowStuffWorks*, 1998–2004, www.science.howstuffworks.com/exoskeleton.htm.

de Bakker, Liesbeth, "The World's Most Energy Efficient Robot," *Radio Netherlands*, Aug. 7, 2000, www.rnw.nl/science/html/robots000807.html.

Dixon, Chuck, and Leonardo Manco, *Doom* (New York: Marvel Books, 2002).

"Dr. Doom," *The Vault*, www.advancediron.com/drdoom.html. An unofficial Marvel Web page with information on various Marvel villains.

Harris, Tom, "How Body Armor Works," *HowStuffWorks*, 1998–2004, www.people.howstuffworks.com/body-armor.htm.

Hillman, Tyler, "Prof. Begins Research on Futuristic Exoskeleton," *Daily Californian*, Jan. 30, 2002, www.dailycal.org/article.asp?id=7517.

Hirshfeld, Bob, "Deploying Exoskeletons," *Tech TV*, Oct. 16, 2002, www.techtv.com/news/scitech/story/0,24195,3371422,00.html.

Lee, Stan, and Jack Kirby, *Marvel Masterworks* #2, *The Fantastic Four* #1–10 (New York: Marvel Books, 1987).

———, *Marvel Masterworks* #6, *The Fantastic Four* #11–20 (New York: Marvel Books, 1988).

———, *Marvel Masterworks* #13 *The Fantastic Four* #21–30 (New York: Marvel Books, 1990).

"Powered Exoskeleton—Legged Vehicle," *Spring Walker, Body Amplifier*, www.springwalker.com/. Information on this new type of exoskeleton device.

Ward, Mark, "The Military Gets Mightier," *BBC News, Sci/Tech*, Jan. 12, 2001, www.news.bbc.co.uk/1/hi/sci/tech/ 1112411.stm.

3. Computer Supervillain or Village Idiot? Brainiac

Adventures of Superman #438, Mar. 1988.

Adventures of Superman #445, Oct. 1988.

"The Brainiac Trilogy," *Action Comics* #649, Jan. 1990.

"The Super-Duel in Space," *Action Comics* #242, 1958.

Superman Y2K (continuing comic book series).

Brooks, Rodney, "Elephants Don't Play Chess," *Robotics and Autonomous Systems* (North Holland: Elsevier Science Publishers, 1990).

———, "New Approaches to Robotics," *Science* Sept. 3, 1991.

Drexler, Eric, *Engines of Creation: The Coming Era of Nanotechnology* (New York: Anchor Books/Doubleday, 1986).

Gresh, Lois H., and Robert Weinberg, *The Computers of Star Trek* (New York: Basic Books, 1999).

Oakes, Chris, "The Year 2020, Explained," July 5, 2000, www.wired.com/news/technology/0,1282,37117,00.html.

4. Feathers and Fury: The Vulture

Amazing Spider-Man, vol. 1, #1, 2, 3.

Brain, Marshall, and Brian Adkins, "Aerodynamic Forces," *HowStuffWorks*, 1998–2004, www.travel.howstuffworks.com/airplane1.htm.

———, "How Airplanes Work," *HowStuffWorks*, 1998–2004, www.travel.howstuffworks.com/airplane2.htm.

Chang, Amelia, "Species," *All About Parrots*, 2003, www.geocities.com/allaboutparrots.

Ditko, Steve, and Stan Lee, *Amazing Fantasy* #15 (New York: Marvel Comics, Aug. 1962).

Ducker, David, "Aliens Explore Earth, Birds," *Alien Explorer*, 2003, www.alienexplorer.com.

Hornby, Nick, "To Fly Like a Bird," *Aviation*, 2000, www.ucl.ac.uk/slade/nickhornby/flying/extra/flahoo.html.

van der Linden, R., "The MacCready Gossamer Condor," National Air and Space Museum, www.nasm.si.edu/aircraft/maccread_condor.htm.

"The Vulture," *Enemies of Spider-Man*, www.alaph.com/spiderman/enemies/vulture.html. Fan page dedicated to the various foes of Spider-Man, with a bibliography of their appearances.

5. The Kiss of Death: Poison Ivy

"Beware of Poison Ivy!" *Batman* #181, June 1966.

Advertisement for poison ivy remedy, www.zanfel.com.

American Academy of Dermatology, "Poison Ivy, Oak & Sumac," *AAD Public Resources*, www.aad.org/pamphlets/PoisonIvy.html.

Armstrong, W. P., and W. L. Epstein, "Poison Oak: More Than Just Scratching the Surface," *Herbalgram*, vol. 34, 1995.

Dulac, Catherine, "Pheromones Control Gender Recognition in Mice," *HHMI News*, Jan. 7, 2002, www.hhmi.org/news/dulac.html.

———, "Pheromones Receptors Need 'Escorts,'" *HHMI News*, Mar. 7, 2003, www.hhmi.org/news/dulac2.html.

Dunphy, Jim, "Urushoil Oil Is Potent," Poison Ivy, Oak, and Sumac Information Center, 1999–2004, www.poisonivy.aesir.com/view/fastfacts.html.

Epstein, W. L., H. Baer, C. R. Dawson, and R. G. Khurana, "Poison Oak Hyposensitization: Evaulation of Purified Urushiol," *Archives of Dermatology*, 1974.

Gross, Michael, *Travels to the Nanoworld: Miniature Machinery in Nature and Technology* (New York: Perseus Publishing, 1999).

Keller, David H., "The Ivy War," *Amazing Stories*, May 1930.

Kelly, James Patrick, "Mr. Boy," *Asimov's Science Fiction Magazine*, June 1990.

Leinster, Murray, "Proxima Centauri," *Astounding Science Fiction*, Mar. 1935.

Long, Frank Belknap, "Plants Must Kill," *Thrilling Wonder Stories*, Apr. 1942.

Martz, Eric, "Poison Ivy: An Exaggerated Immune Response to Nothing Much, Mar. 31, 1997, www.bio.umass.edu/immunology/poisoniv.htm.

Rudolph, Frederick B., and Larry V. McIntire (eds.), *Biotechnology: Science, Engineering, and Ethical Challenges for the 21st Century* (Washington, D.C.: Joseph Henry Press, 1996).

Stehlin, Isadora, B., "Outsmarting Poison Ivy and Its Cousins," *FDA Consumer Magazine*, Sept. 1996, www.fda.gov/fdac/features/796_ivy.html.

Vance, Jack, "The Houses of Iszm," *Startling Stories*, spring 1954.

World Health Organization, *Chloroform Health and Safety Guide* (Geneva: World Health Organization for the International Programme on Chemical Safety, 1994).

6. Groping for Power: Doctor Octopus

"Disaster!" *Marvel Tales* #41, Jan. 1973.

"Doctor Octopus," *Marvel Tales* #38, Oct. 1972.

"Spider-Man versus Doctor Octopus," *Amazing Spider-Man*, vol. 1, #3, 1963.

Anderson, Mark K., "Becoming Your Own Hospital," *Wired News*, Nov. 11, 2000, www.wirednews.com/news/print/0,1294,40120,00.html.

———, "Mega Steps Toward the Nanochip," *Wired*, Apr. 27, 2001.

Biomimetric.com, "Frequently Asked Questions About Biomimetrics and Smart Materials," 1997–2002, www.biomimetic.com/faq.html.

Drexler, K. Eric, *Engines of Creation* (Garden City, N.Y.: Anchor Books, 1986).

———, "Ask the Chemistry Expert, Machine-Phase Nanotechnology,"

ScientificAmerican.com, Sept. 16, 2001, www.sciam.com/askexpert/
chemistry/chemistry6.html.

Gresh, Lois H., *TechnoLife 2020: A Day in the World of Tomorrow* (Toronto,
Canada: ECW Press, 2001).

Gross, Michael, *Travels to the Nanoworld: Miniature Machinery in Nature and
Technology* (New York: Perseus Publishing, 1999).

Nighswonger, Gregg, "New Polymers and Nanotubes Add Muscle to Prosthetic Limbs," *Medical Device & Diagnostic Industry*, Aug. 1999, as
reported at www.devicelink.com/mddi/archive/99/08/004.html.

O'Malley, Chris, "The Binary Man: Step One," *Popular Science*, Mar. 1999.

"Smart Materials," *Scientific American*, May 1996.

"Smart Materials and Structures at Sandia," *Sandia National Laboratories*, 2003,
www.endo.sandia.gov/9234/smas.html.

Stix, Gary, "Ask the Chemistry Expert, Little Big Science," ScientificAmerican
.com, Sept. 16, 2001, www.sciam.com/askexpert/chemistry/chemistry6
.html.

Swissler, Mary Ann, "Microchips to Monitor Meds," *Wired News*, Sept. 28,
2000, www.wirednews.com/news/print/0,1294,39070,00.html.

"Tech 2010," *New York Times Magazine*, June 11, 2000.

Zey, Michael G., *The Future Factor: The Five Forces Transforming Our Lives and
Shaping Human Destiny* (New York: McGraw-Hill, 2000).

www.marvel.com, the home page for Marvel Comics, featuring biographies of
some of their most famous heroes and villains, including Dr. Octopus.

www.oandp.com/news/, the O&P Edge, the Web site for cutting-edge information on prosthetics, orthotics, and allied health care professionals.

7. Leapin' Lizards: The Lizard

Amazing Spiderman #6, 1963; #32, 1965; #33, #43, #44, 1966; #45, 1967.

Bryant, Susan, and David Gardiner, "Axolotol Limb Regeneration," *LegLab
UC Irvine*, www.darwin.bio.uci.edu/~mrjc/Regeneration/complete.html.

Craig, Owen, "Spinal Repair," *Quantum TV*, Apr. 5, 2001, www.abc.net.au/
quantum/s267523.htm.

Gordon, Serena, "Heart Regeneration Proved Possible," *Health on the Net
Foundation*, Dec. 12, 2002, www.hon.ch/News/HSN/510777.html.

LeBold, Anne, "What Stimulates Limb Regeneration," *Research*, summer
1995, www.uoguelph.ca/research/publications/Assets.html.

"Man Is Turning Himself into Lizard," *Ananova*, July 23, 2003, www.ananova
.com/news/story/sm_290461.html.

Sahm, Phil, "The Mystery of Regeneration," *University of Utah Health Sciences Report*, summer 2003, www.uuhsc.utah.edu/pubaffairs/hsr/summer2003/regeneration.html.

"Skeletal Muscle Regeneration," www.angelfire.com/me3/Schallek/documents/bioengineering. A detailed study of how skeletal muscle fibers regenerate and why this subject is important to human healing.

"Spinal Repair," *Quantum Television*, Aug. 18, 2003, www.abc.net.au/quantum/s267523.htm.

Travis, John, "Starting Over," *Science News Online, This Week*, Nov. 11, 1997.

Welsome, Eileen, *The Plutonium Files* (New York: Bantam Books, 1999).

8. Clothes Make the Man: Venom

Amazing Spider-Man #252–258, 1984.

Amazing Spider-Man #298–300, 1988; #315–317, 1989; #330–333, 1990; #345–347, 1991; #362–363, 1992; #373–375, 1993.

Amazing Spider-Man Annual #25, 1991; 26, 1992.

Maximum Carnage (New York: Marvel Books, 1995).

Secret Wars #8, 1983.

Spectacular Spider-Man Annual #12, 1992.

Venom: Carnage Unleashed #1–4, 1995.

Venom: The Enemy Within #1–4, 1994.

Venom: Funeral Pyre #1–4, 1993.

Venom: The Hunted #1–3, 1996.

Venom: The Knights of Vengeance #1–4, 1994.

Venom: Lethal Protector #1–6, 1993.

Venom: The Madness #1–4, 1993–1994.

Venom: Separation Anxiety #1–4, 1994–1995.

Web of Spider-Man #1, 1985; #18, 1986.

Web of Spider-Man Annual #8, 1992.

Chai, Winston, "Smart-Clothes Chips End Up in Dirty Laundry," *ZDNet*, Apr. 7, 2003, www.zdnet.com/2102-1105-995722.html.

Clark, Ken, "'Smart' Clothes Feature Sensors and Cameras," *TDC Trade.Com*, March 2000, www.tdctrade.com/imn/imn177/gallery.htm.

Inami, Masahiko, and Naoki Kawakami, "Optical Camouflage," *Tokyo Star*, May 2003, www.star.t.u-tokyo.ac.jp/projects/MEDIA/xv/oc.html.

International Center of Excellence for Wearable Electronics and Smart Clothing, *ICEWES Overview*, www.icewes.net/projdetails.htm. A detailed report on research involving smart clothing and the European Union effort to be a major producer of such clothes in the near future.

Pentland, A., "It's Alive," *IBM Systems Journal*, vol. 39, #3, 4 (2000), www .research.ibm.com/journal/sj/393/part3/pentland.html.

Schowengerdt, Richard Neal, *Project Chameleo*, Nov. 23, 2003, www .chameleo.net/project.html.

"The Smart Jacket," *Discovery.Com News*, Dec. 16, 2002, www.disc.discovery .com/news/briefs/20021216/smartjacket.html.

Valigra, Lori, "Fabricating the Future," *Christian Science Monitor, Sci/Tech*, Aug. 29, 2002, www.csmonitor.com/2002/0829/p11s01-stgn.html.

"Who Is Venom?" www.members.aol.com/venomx007%20/whoisvenom .html. An unofficial Marvel Comics fan page dealing with the character, Venom, and his many appearances in the Spider-Man comic book series.

www.icewes.net/HNW.htm. Evonetics, another development in the smart clothing field by the European Union and the use of high-tech fashion technology.

9. Grodd Almighty: The Evil Super-Gorilla

The Flash #106, April–May 1959.

"Return of the Super-Gorilla" *The Flash* #107, June 1959.

Anderson, Poul, *Brain Wave* (New York: Ballantine Books, 1954).

Boysen, Sally, "Chimps R Us—What Are They Thinking?" *PBS, Scientific American Frontiers*, www.pbs.org/saf/1108/features/boysen.htm.

Calvin, Willilam H., "The Emergence of Intelligence," *Scientific American*, Oct. 1994.

"A Conversation with Koko the Gorilla," *PBS, Nature*, www.pbs.org.wnet/ nature/koko.

Deacon, T. W. "Fallacies of Progression in Theories of Brain-Size Evolution," *International Journal of Primatology*, 1990.

10. A Magnetic Personality: Magneto

Magneto has made so many guest appearances in not only the X-Men comics but in all of the Marvel superhero comic books that giving even an abbreviated list of all of his appearances would fill pages. A sampling of his many battles with the X-Men is included in this brief survey.

Avengers #47–53.

Fantastic Four #102–104.

Magneto: Dark Seduction #1–4.

Magneto Rex #1–3, 1999.

Uncanny X-Men #1–45, 62–103.

Uncanny X-Men #366–367, 1999.

X-Men #85–87, 1999.

X-Men: The Magneto War (crossover event), 1999.

AntiGravity Propulsion, A Faster Way to the Stars, April 2001, www .antigravitypower.tripod.com/magnetism/.

Barrett, Stephen, "FAQs," *Quackwatch*, Aug. 9, 2003, www. quackwatch.org/.

Carroll, Robert Todd, "Magnet Therapy," *The Skeptic's Dictionary*, Dec. 10, 2003, www.skepdic.com/magnetic.html.

Daviel, Andrew, "The Basic Forces," *The Triumf Cybertour*, May 2000, www.sundae,triumf.ca/pub2/ctour/pamphlets/forces.html.

"Electricity," *Encarta*, copyright 1993–2004 by Microsoft Co., www.encarta .msn.com/encnet/refpages/RefArticle.aspx?refid=761566543.

"Four Basic Forces" *Science Park*, Particle Physics, 2001, www.science-park .info/particle/forces.html.

"The Four Fundamental Forces of Nature," *The Modular Approach to Physics*, University of Calgary, www.kingsu.ab.ca/~brian/meltp/edit_content/ explainit/forces/FFOUR/four_t1.html.

"The Frog That Learned to Fly," HFML, University of Nijmegen, July 10, 2000, www.hfml.kun.nl/froglev.html.

Geim, Andrey, "Everyone's Magnetism," *Physics Today*, Sept. 1998, www .hfml.kun.nl/phystod.html.

Gresh, Lois, and Robert Weinberg, *The Science of Superheroes* (Hoboken, N.J.: Wiley, 2002).

Joo, John, Jerome Kare, Derek Shiell, and Jeremy Watchuka, "Electromagnetic Weapons," Northwestern University, 2001, www.physics.northwestern .edu/classes/2001Fall/Phyx135–2/19/emp.htm.

Kurtus, Ron, "Succeed in Physics—Forces," School for Champions, Kurtus Technologies, Oct. 2003, www.school-for-champions.com/science/.

"Magneto FAQ," www.geocities.com/Area51/Keep/4026/MagnetoBio.html. A fan biography of Magneto based on information gathered from numerous Marvel comics appearances.

"New Invention Allows Humans to Live Forever," *AlexChiu.com*, www.alexchiu .com/eternallife/.

Pierson, Laura, and Stella Kurtz, "A Description of Magneto's Powers," www
.alara.net/xbooks/magdesc.txt.

Smith, George, "Department of Defense Employees Hold Forth on Radio
Frequency Weapons for Congress," *Crypt Newsletter 47*, 1997, www
.soci.niu.edu/~crypt/other/house.htm. The foremost source of information and truth about cyberterrorism on the Internet.

Stern, David P., and Mauricio Peredo, "Magnetism," *The Exploration of Earth's
Magnetosphere*, Jan. 1, 2003, www-istp.gsfc.nasa.gov/Education/Imagnet
.html.

11. Immortality for Some: Vandal Savage and Apocalypse

Action Comics #515, #516, 1981; #542, #543, 1983; #552, #553, #556, 1984.

All-Star Comics #37, 1947; #64, #65, 1977.

DC One Million #1, 2, 3, 1998.

DC Universe Villains Secret Files #1, 1999.

Flash, vol. 1, #137, 1963; #215, 1972; #235, 1975.

Flash, vol. 2, #1, #2, 1987; #48, #49, #50, 1991; #124, 1997.

The Further Adventures of Cyclops and Phoenix #1–4, 1996.

Green Lantern, vol. 1, #10, 1943.

Justice League of America #88, 1994.

Justice League Task Force #17, #18, #19, #20, #21, 1994–1995.

Justice Society of America #1–8, 1991.

The Rise of Apocalypse #1–4, 1996.

Titans #5–12, 1999–2000.

X-Factor #5, #6, 1986; #24, 1988.

X-Force #37, 1994.

Bailey, Ronald, "Forever Young," *Reason Online*, August 2002.

Bradbury, Robert, *Lifespan Extension Issues*, www.aeiveos.com/issues.html.

Haigh, Richard, and Mirko Bagaric, "Immortality and Sentencing Law," *Journal of Philosophy, Science and Law*, May 2002, www.psljournal.com/
archives/papers/immortality.cfm.

"Immortality Becomes Real," *Pravda*, July 2, 2003.

Roach, Kate, "Immortality—Hype or Hope?" *Science in Medicine*, May 2003,
www.channel4.com/science/microsites/S/science/medicine/liveforever
.html.

Wikipedia, the Free Encyclopedia: Retcon, http://en.wikipedia.org/wiki/Rotcon.

12. Have Surfboard, Will Travel: The Silver Surfer

Fantastic Four, #48–50, 55, March, April, May, Nov. 1966.

Super-Villain Classics #1, May 1983.

Thor #161, Feb. 1969.

Durda, Dan, and Mark Gelfand, "Solar Sailing," *The Planetary Society*, Sept. 19, 2003, www.planetary.org.

Gresh, Lois, and Robert Weinberg, *The Science of Superheroes* (Hoboken, N.J.: Wiley, 2002).

Hydarai, Ezat, *Solar Winds as a Possible Cause of the Permian-Triassic Boundary Extinction*, Jackson State University, Nov. 6, 2001, www.gsa.confex .com/gsa/ 2001AM/finalprogram/abstract_21288.htm.

Jackson, Deborah J., "Solar Sail Technology," *Jet Propulsion Laboratory*, Apr. 29, 2002, www.solarsails.jpl.nasa.gov/introduction/.

Lee, Stan, and John Buscema, *Marvel Masterworks: The Silver Surfer* (New York: Marvel Books, 1990).

———, *Marvel Masterworks: The Silver Surfer 2* (New York: Marvel Books, 1991).

Pedrick, James, "Galactus, the Webpage," 2001, www.marvelite.prohost-ing.com/surfer/galactus/. An authorized Marvel Web site.

"Sailing to the Stars," *In the News*, Glencoe Online, Apr. 11, 2001, www.glencoe. com/sec/science/physics/wwwlinks/updates/archives/ch6sail.php?yr=Physics.

Simberg, Rand, "Sailing, Sailing," *Transterrestrial Musings*, May 14, 2002, www.interglobal.org/weblog/archives/001036.html.

13. The Case of the Missing Antimatter: Sinestro

Green Lantern (Silver Age) #7, #9, 1961; #11, #15, 1962; #18, 1963; #52, 1967; #59, 1968; #73, 1969; #74, 1970; #82, 1971; #91, 1976; #92, 1977; #123–125, 1979; #127, 1980; #197–198, 1986; #200, 1986; #217–224, 1987–1988.

Green Lantern—Emerald Dawn #5, #6, 1990.

Green Lantern—Emerald Dawn vol. 2, #1–6, 1991, as well as in numerous other Green Lantern Corps adventures.

Calder, Neil, "From Theory to Certainty, BaBar Announces New Result on Charge Parity Violation," *Eureka Alert!* Doe/Stanford Linear Accelerator Center, July 23, 2002, www.eurekalert.org/pub_releases/2002–07/ dlac-ftt082103.php.

Casper, Dave, "Super-Kamiokan at UC Irvine," July 1998, www.ps.uci.edu/ ~superk/index.htm.

Cohen, A. G., A. de Rugula, and S. L. Glashow, *A Matter-Antimatter Universe?* Nov. 15, 1997, www.arxiv.org/abs/astro-ph/9707087.

de Rujula, Alvaro, and Rolf Landua, "Antimatter: Mirror of the Universe," *Live from Cern*, 2000–2001, www.livefromcern.web.cern.ch/livefromcern/ antimatter/.

Gresh, Lois, and Robert Weinberg, *The Science of Superheroes* (Hoboken, N.J.: Wiley, 2002).

Grimes, Jack, and Michael Bond, "Emerald Genesis, The History of the Green Lantern Corps," April 23, 2000, www.glcorps.org/_genesis.html. The unofficial Green Lantern Corps page.

Kaku, Michio, "What Happened Before the Big Bang," *Cosmology Today*, Jan. 1, 1998, www.flash.net/~csmith0/bigbang.htm.

Katayama, Fred, "Antimatter Could Fuel Rockets, Heal Patients," *CNN .Com/Space*, Jan. 10, 2002, www.cnn.com/2002/TECH/space/01/10/ antimatter.research/.

Moyer, Michael, "Antimatter," *Popular Science*, 2004, www.popsci.com/popsci/ science/article/0,12543,220659-4,00.html.

Particle Data Group, "The Scale of the Atom," *The Particle Adventure*, 2002, www.particleadventure.org/particleadventure/frameless/scale.html.

———, "The Standard Model," *The Particle Adventure*, 2002, www .particleadventure.org/particleadventure/frameless/standard_model. html.

Whitehouse, David, "Big Balloon Bags Antimatter," *BBC News, Sci/Tech*, Aug. 18, 1999, www.news.bbc.co.uk/1/hi/sci/tech/423656.stm.

14: Crisis on Infinite Earths

Bellevue Community College, "Cantor's Diagonalization Argument," 1999–2003, www.scidiv.bcc.ctc.edu/Math/diag.html.

———, "Counting to Infinity," 1999–2003, www.scidiv.bcc.ctc.edu/Math/ infinity.html.

Casey, Nancy, and Mike Fellows, "One-to-One Correspondence," *Mega-Math*, Los Alamos National Laboratory, 1995, www.c3.lanl.gov/mega-math/gloss/infinity/onetoone.html.

———, "Welcome to Hotel Infinity," *Mega-Math*, Los Alamos National Laboratory, 1991, www.c3.lanl.gov/mega-math/workbk/infinity.

Fox, Gardner, and Carmine Infantino, "Flash of Two Worlds," *Flash Comics* #123, Sept. 1962.

Math Academy Online, "Cantor's Theorem," *Platonic Realms Interactive Mathematics Encyclopedia*, 1997–2004, www.mathacademy.com/pr/prime/articles/cantor_theorem/.

St. Andrews School of Mathematics and Statistics, "Cantor's Diagonal Argument," May 27, 1999, www.shef.ac.uk/~puremath/theorems/cantor.html.

Wolfman, Marv, and George Perez, *Crisis on Infinite Earths* (New York: DC Comics, 1985, 1998).

15. Frustration in Five Dimensions: Mr. Mxyzptlk and Bat-Mite

"Batman Meets Bat-Mite," *Detective Comics* #267, May 1959.

"Bat-Mite Meets Mr. Mxyzptlk," *World's Finest Comics* #133, Nov. 1960.

"The Incredible Team of Bat-Mite and Mr. Mxyzptlk," *World's Finest Comics* #143, Feb. 1962.

"Mr. Mxyzptlk Returns," *Action Comics* #80, Jan. 1945.

"The Mysterious Mr. Mxyzptlk," *Superman* #30, Sept./Oct. 1944.

Banshee, MuTed, "Mr. Mxyzptlk," Cellar of Comics, 2003, www.cocs.com/mrmxyzptlk.asp.

Bara, Mike, "A Unique Opportunity for Testing the Exploding Planet Hypothesis and Hyperdimensional Physics," *The Enterprise Mission*, www.enterprisemission.com/hyper1a.html.

Burger, Dionys, *Sphereland*, trans. C. J. Rheinboldt (New York: Thomas Crowell, 1964).

Dorkin, Evan, and friends, "Last Imp Standing," *World's Funnest* (New York: DC Comics, 2000).

Younis, Steven, "Mild Mannered Reviews, Pre-Crisis Reviews," Superman home page, 1996–2004, www.supermanhomepage.com/comics/pre-crisis-reviews/.

Acknowledgments

The authors would like to thank Chris Claremont, Eric Moreels, Andrew Pepoy, Jennifer Contino, Kurt Busiek, Richard Clark, and Jeff Mariotte for their help in preparing this book. Thanks as always to our agent, Lori Perkins, and to our editor at John Wiley & Sons, Stephen S. Power.

Index